T0191985

EAI/Springer Innovations in Communication and Computing

Series Editor

Imrich Chlamtac, European Alliance for Innovation, Ghent, Belgium

Editor's Note

The impact of information technologies is creating a new world yet not fully understood. The extent and speed of economic, life style and social changes already perceived in everyday life is hard to estimate without understanding the technological driving forces behind it. This series presents contributed volumes featuring the latest research and development in the various information engineering technologies that play a key role in this process.

The range of topics, focusing primarily on communications and computing engineering include, but are not limited to, wireless networks; mobile communication; design and learning; gaming; interaction; e-health and pervasive healthcare; energy management; smart grids; internet of things; cognitive radio networks; computation; cloud computing; ubiquitous connectivity, and in mode general smart living, smart cities, Internet of Things and more. The series publishes a combination of expanded papers selected from hosted and sponsored European Alliance for Innovation (EAI) conferences that present cutting edge, global research as well as provide new perspectives on traditional related engineering fields. This content, complemented with open calls for contribution of book titles and individual chapters, together maintain Springer's and EAI's high standards of academic excellence. The audience for the books consists of researchers, industry professionals, advanced level students as well as practitioners in related fields of activity include information and communication specialists, security experts, economists, urban planners, doctors, and in general representatives in all those walks of life affected ad contributing to the information revolution.

Indexing: This series is indexed in Scopus, Ei Compendex, and zbMATH.

About EAI

EAI is a grassroots member organization initiated through cooperation between businesses, public, private and government organizations to address the global challenges of Europe's future competitiveness and link the European Research community with its counterparts around the globe. EAI reaches out to hundreds of thousands of individual subscribers on all continents and collaborates with an institutional member base including Fortune 500 companies, government organizations, and educational institutions, provide a free research and innovation platform.

Through its open free membership model EAI promotes a new research and innovation culture based on collaboration, connectivity and recognition of excellence by community.

More information about this series at http://www.springer.com/series/15427

Akashdeep Bhardwaj • Varun Sapra
Editors

Security Incidents & Response Against Cyber Attacks

 Springer

Editors
Akashdeep Bhardwaj
School of Computer Science
University of Petroleum & Energy Studies
Dehradun, India

Varun Sapra
School of Computer Science
University of Petroleum & Energy Studies
Dehradun, India

Series Editor
Imrich Chlamtac

ISSN 2522-8595   ISSN 2522-8609   (electronic)
EAI/Springer Innovations in Communication and Computing
ISBN 978-3-030-69176-9   ISBN 978-3-030-69174-5   (eBook)
https://doi.org/10.1007/978-3-030-69174-5

This Springer imprint is published by the registered company Springer Nature Switzerland AG
The registered company address is: Gewerbestrasse 11, 6330 Cham, Switzerland

To Those Who Inspired This Book But Will Never Read It!

Foreword

Introduction

The year 2020 has seen a dramatic increase in the number of security incidents and cyberattacks. The COVID-19 pandemic mandated social distancing and working from home. This involuntary digital transformation resulted in a substantial increase in online activities, employee interactions, e-learning, and financial transactions. This led to more opportunities for cybercriminals as the attack surface area became larger than ever. As the community adjusts to the "new normal," there is a good indication that many of us that are working from home will continue to do so and we will never ever get back to "normal." The proliferation of mobile devices and ubiquitous Internet connectivity has also increased the threat surface area. The digital transformation of our lives, our work, and our social activities is likely to result in more incidents and attacks. Therefore, preparing, planning, and mitigating security incidents and responses to the cyberattacks must become part of an organization and every individuals' daily routines in the "new normal."

Cybersecurity incidents and cyberattacks have been around for as long as computers have been around, but the complexity and sophistication of these incidents and attacks recently, the magnitude of attacks, costs, and irreparable damage are something that cannot be ignored anymore. Security incident planning and mitigation is critical to an organization survival. Being able to counter cyberattacks and prevent attacks must be the requirement of an organization's standard operating procedures. Part of the blame for the increase in security incidents and cyberattacks lies with organization, institutions, and individuals trying to prevent them. For example, organizations like Cisco provide comprehensive training about computer networks and cybersecurity. This provides in-depth knowledge of the networks/protocols and how to exploit them. Institutions teach courses on ethical hacking and individuals make and upload "how to hack" videos on YouTube.

To make matters worse, security incidents and cyberattacks are now being armed with technologies like Artificial Intelligence, which can easily outsmart human responses in trying to detect, mitigate, and counter attacks. There is a need to train more cybersecurity analysts and equip them with the state-of-the-art hardware and software to mitigate future incidents and attacks. Virtual Private Networks, the TOR browser, and the Dark Net are some of the tools among many others used by hackers, thus making it difficult to identify the perpetrators. Cybersecurity has remained and will remain a "cat-and-mouse" game. There will be attacks and exploits and then there will be patches, upgrades, and antiviruses. We will always remain one step behind the attackers. Our best effort in this game is to have proper procedures in place, the hardware and the software to detect security incidents and cyberattacks and mitigate them.

Along the way, we have learnt a lot and developed ISO27000 standards, NIST 800 61 Incident Handling Guides, as well as local and regional Computer Security Incident Response Teams (CSIRT). CERT allows constituents to report information via phone, email, or by using a secure incident-reporting website. They accept reports of security incidents, phishing attempts, malware, and vulnerability reporting. Standards and guidelines are adopted and followed to mitigate security incidents and cyberattacks. These standards and guidelines are best practices that have been developed over years for organizations to adopt. They are techniques generally set forth in published materials that attempt to protect the cyber environment of a user or organization. Security certified and accredited organizations find it easier to collaborate their networks.

Security Incidents

A security incident or cyberattack takes place when there is an unauthorized access to an organization's computer network. This access could be to get information, compromise the integrity of data, or make the system unavailable. This is an incident or attack that breaches the CIA triad of confidentiality, integrity, and availability of the system. Most organizations will have a cybersecurity team or a network systems administrator to monitor their networks, detect, and mitigate such incidents and attacks. According to Rosencrance (2019) "Security incidents are events that may indicate that an organization's systems or data have been compromised or that measures put in place to protect them have failed. In IT, a security event is anything that has significance for system hardware or software, and an incident is an event that disrupts normal operations. A security breach is a confirmed incident in which sensitive, confidential, or otherwise protected data has been accessed or disclosed in an unauthorized fashion."

The University of California Berkeley Information Security Office describes a security incident as "an event that leads to a violation of an organization's security policies and puts sensitive data at risk of exposure." Security incident is a broad term that includes many kinds of events.

According to SANS (Pokladnik 2020), there are six key phases of an Incident Response Plan:

1. Preparation: Preparing users and IT to handle potential incidents in case they happen (and let's face it, we know they will)
2. Eradication: Finding and eliminating the root cause (removing affected systems from production)
3. Identification: Figuring out what we mean by a "security incident" (which events can we ignore vs. which we must act on right now?)
4. Recovery: Permitting affected systems back into the production environment (and watching them closely)
5. Containment: Isolating affected systems to prevent further damage (automated quarantines are our favorite)
6. Lessons Learned: Writing everything down and reviewing and analyzing with all team members so you can improve future efforts

   Examples of security incidents include:

- Computer system breach
- Unauthorized access to, or use of, systems, software, or data
- Unauthorized changes to systems, software, or data
- Loss or theft of equipment storing institutional data
- Denial of service attack
- Interference with the intended use of IT resources
- Compromised user accounts

It is important that actual or suspected security incidents be reported as early as possible so that organizations can limit the damage and cost of recovery.

CERT and CSIRT

Computer Emergency Response Teams (CERT) and Computer Security Incident Response Teams (CSIRT) are usually responsible for attending to security incidents and cyberattacks at the organizational, national, and regional level. These teams are made up of cybersecurity, information security, and other specialists that are deployed to detect, mitigate, and provide defense in depth security against any further incidents or attacks. A CERT is a group of information security experts responsible for the protection against, detection of, and response to an organization's cybersecurity incidents (Rouse 2019).

It is now common to find CERT/CSIRT at national and regional levels. Almost every country has its own CERT or access to a regional CERT. This indicates the level of response required for security incidents and cyberattacks. CERTs are responsible for coordinating the cybersecurity information that affects every government agency, business, and individual computer user within their jurisdiction. They provide security alerts, vulnerability information, and helpful tips for protecting an organization or a home user. To be effective, CERT needs to receive security incidents from its users. CERT allows constituents to report information via phone, email, or by using a secure incident-reporting website. They accept reports of security incidents, phishing attempts, malware, and vulnerability reporting.

Standards and Guidelines

Standards and guidelines are adopted and followed to mitigate security incidents and cyberattacks. These standards and guidelines are best practices that have been developed over years for organizations to adopt. They are techniques generally set forth in published materials that attempt to protect the cyber environment of a user or organization. This environment includes users themselves, networks, devices, all software, processes, information in storage or transit, applications, services, and systems that can be connected directly or indirectly to networks. The principal objective is to reduce the risks, including prevention or mitigation of cyberattacks. These published materials consist of collections of tools, policies, security concepts, security safeguards, guidelines, risk management approaches, actions, training, best practices, assurance, and technologies.

For example, the ISO (International Organization for Standardization)/IEC (International Electro technical Commission) 27001 and 27002 standards formally specify a management system that is intended to bring information security under explicit management control. ISO/IEC 27001 is an international standard on how to manage information security. It details requirements for establishing, implementing, maintaining, and continually improving an information security management system (ISMS)—the aim of which is to help organizations make the information assets they hold more secure. ISO/IEC 27002 is an information security standard that provides the best practice recommendations on information security controls for use by those responsible for initiating, implementing, or maintaining information security management systems (ISMS). The NIST (National Institute for Standards and Technologies) Cybersecurity Framework has a number of publications for guidelines. For example: Special Publication 800-12 provides a broad overview of computer security and control areas. Special Publication 800-14 describes common security principles that are used. Special Publication 800-53 provides information on how to manage IT security.

Cyberattacks

Cybercrime is the greatest threat to every company in the world, and one of the biggest problems with humanity. The impact on society is reflected in the numbers. In 2016, Cybersecurity Ventures predicted that cybercrime would cost the world $6 trillion annually by 2021, up from $3 trillion in 2015. This represents the greatest transfer of economic wealth in history, risks the incentives for innovation and investment, and will be more profitable than the global trade of all major illegal drugs combined (Morgan 2017). Global spending on cybersecurity will exceed $1 trillion cumulatively over the next five years, according to Cybersecurity Ventures. Cybersecurity Ventures predicts that a business will fall victim to a ransomware attack every 14 seconds by 2019. According to Cybersecurity Ventures, training employees on how to recognize and react to phishing emails and cyber threats may be the best security Return on Investments.

Writing in Accenture's Ninth Annual Cost of Cybercrime study, Bissell et al. (2019) state "Cyberattacks are evolving from the perspective of what they target, how they impact organizations and the changing methods of attack." They found that cyberattacks are changing due to evolving targets, evolving impact, and evolving techniques. The expanding threat landscape and new business innovation are leading to an increase in cyberattacks. The average number of security breaches in the last year grew by 11% from 130 to 145. Organizations spend more than ever to deal with the costs and consequences of more sophisticated attacks. The average cost of cybercrime for an organization increased from US$1.4 million to US$13.0 million.

Response Against Cyberattacks

Cyberattacks are evolving from the perspective of what they target, how they affect organizations, and the changing methods of attack. Improving cybersecurity protection can unlock economic value by reducing the cost of cybercrime and opening up new revenue opportunities. By understanding where they can gain value in their cybersecurity efforts, leaders can minimize the consequences and even prevent future attacks. By prioritizing technologies that improve cybersecurity protection, organizations can reduce the consequences of cybercrime and unlock future economic value as higher levels of trust encourage more business from customers. Improving cybersecurity protection can decrease the cost of cybercrime and open up new revenue opportunities (Bissell et al. 2019).

According to Bissell et al. (2019), the three steps to unlocking the value in cybersecurity are as follows:

1. Priorities protecting people-based attacks: Countering internal threats is still one of the biggest challenges with a rise in phishing and ransomware attacks, as well as malicious insiders.
2. Invest to limit information loss and business disruption: Already the most expensive consequence of cyberattacks, this is a growing concern with new privacy regulations such as GDPR and CCPA.
3. Target technologies that reduce rising costs: Use automation, advanced analytics, and security intelligence to manage the rising cost of discovering attacks, which is the largest component of spending.

Organization of the Book

The book is organized into eleven chapters. A brief description of some of the chapters follows.

Foreword

This introduction chapter presents the different concepts and context of the book. It explains how the pandemic has led to digital transformation and therefore a dramatic increase in security incidents and cyberattacks. How training to mitigate security incidents and cyberattacks has resulted in more hackers and cybercriminals being trained. To make matters worse, security incidents and cyberattacks are now being armed with technologies like Artificial Intelligence, which can easily outsmart human responses in trying to detect, mitigate, and counter attacks. Along the way, we have learnt a lot and developed ISO27000 standards, NIST 800 61 Incident Handling Guides, as well as local and regional Computer Security Incident Response Teams (CSIRT). A security incident or cyberattack takes place when there is an unauthorized access to an organization computer network. Computer Emergency Response Teams (CERT) and Computer Security Incident Response Teams (CSIRT) are usually responsible for attending to security incidents and cyberattacks at the organizational, national, and regional level. It is important that actual or suspected security incidents be reported as early as possible so that organizations can limit the damage and cost of recovery.

Chapter 1: By Failing to Prepare, You Are Preparing to Fail

This chapter addresses the critical need to be prepared to respond to incidents and events that may cause a business disruption. It includes benign activities like policy and procedure development and active activities like training, awareness, drills, solutions, automation and more—small components that make up the whole in respect of the domain of Incident Response and Management, as part of the Cybersecurity Management System in an organization. Unfortunately, organizations do not provision for establishing a separate IR/IM function, and this is a weakness in the overall management system. Options are available to outsource IR/IM expertise and the security office must exercise diligence when seeking to contract the function. The author will provide guidance to set up as well as pointers to remember if taking the outsourcing route. Nothing underscores the importance of planning and preparing as the saying "By failing to prepare, you are preparing to fail."

Chapter 2: Design of Block-Chain Polynomial Digests for Secure Message Authentication

The advent of the internet and cloud solutions has completely transformed the conventional storage and retrieval mechanisms. In modern computing, the data are not under the manual control of the owner. This demands radical solutions to be incorporated for this model to address the integrity violations of the remote data. This chapter attempts to perform the functional analysis of standard digest functions and it proposes an idea for the design of a block-chain-based 512-bit digest function using polynomials. Besides, this chapter attempts to examine the erratic behavior of the proposed design through the avalanche response, near-collision response, and statistical analysis of confusion and diffusion. The result proves the response of the proposed design is random and the proposed design meets the strict avalanche criteria. Therefore, the block-chain-based polynomial digest could be considered as an alternative for the contemporary digest function in the perspective of security.

Chapter 3: Collaborative Approaches for Security of Cloud and Knowledge Management Systems: Benefits and Risks

Cloud computing pattern is becoming more and more trendy, due to the enormous decrease in the time, cost, and effort to conversation software program development needs. A knowledge management approach is highly utilized in enterprises in appliances such as intelligence methods, employment authority, and supercomputer learning, in the public domain, protection, and conventional administration. Knowledge management remains believed towards remaining a few of these Categories issue parts. To identify, develop, describe, repository, or disseminate knowledge, the existing devices, the equipment, or techniques that must not be stayed knowledgeable to achieve something the agreement needs always-needed intended used for growth.

Chapter 4: Exploring Potential of Transfer Deep Learning for Malicious Android Applications Detection

Mal Image represents any type of malicious executable (Windows files, APKs) for using image-based features for building classifiers. With the advancement in computing capacities in recent years, the deep learning-based image classification techniques report very high accuracy for different classification tasks such as face detection and recognition, object identification, etc. In this chapter, the authors have combined these two evolving techniques to improve Android malware detection. For this chapter, the research involved experiments with Transfer Learning

techniques under Deep Learning models and Android Malware detection techniques. The experimental result of various pre-trained models in terms of accuracy
is in the range of 75–80%, but this technique can overcome bottlenecks such as
analysis obstacles and obfuscation of traditional methods.

Chapter 5: Exploring and Analysing Surface, Deep, Dark Web, and Attacks

Data is termed as a huge asset in today's world. In this chapter, an introduction to
WWW, classification of different kinds of web, i.e., surface web, deep web, and
dark web, is discussed along with differences among them. Trending research on
deep and dark web is discussed focusing on benefits of deep web. The significance
of searching deep web data underneath the Surface Web aids in getting access to
gigantic data as 96% of data is hidden inside the deep web and it is freely available.
TOR is a tool to access the deep data, and how this works along with its benefits is
deliberated and is the objective of this chapter. Deep web accessing method is
described in detail with suitable examples. Ongoing research in deep web is discussed, and later, attacks faced by the deep web and how cybercriminals use the
dark web are emphasized. An overview of the web, types of the web, and how it
works are discussed focusing on surface web, deep web, and dark web. Distinguishing
characteristics between the deep and dark web are portrayed well with suitable
examples. Attacks faced by the deep web are explained and the need to secure individuals system when accessing data hidden deeply inside the web, and necessary
measures to be considered are discussed.

Chapter 6: Securing ERP Cyber Systems by Preventing Holistic Industrial Intrusion

Organizations are storing various financial and operation data into the ERP system.
Hence security of the ERP systems has become a major concern for organizations.
The situation is becoming more intense after the introduction of IOT wherein organizations are connecting multiple devices linked to the network to control various
aspects of the business. Moving to connected IP systems not only provides the automation because existing systems require expensive maintenance and are complex.
ERP systems are proprietary software, which were made to be used inside the four
walls of an enterprise and are more prone to cyberattacks. Traditionally, ERP systems have been looking at security from responsibilities, role-based access, and
segregation of duties perspective. Hence, a framework or techniques are needed to
stop the malicious attacks on the ERP system. This chapter discusses the few

architecture frameworks to secure ERP using the current infrastructure of any enterprise and provide a secure system at no extra cost.

Chapter 7: Infrastructure Design to Secure Cloud Environments Against DDoS-Based Attacks

This chapter focuses on the creation of a detailed and stable networking architecture to minimize the attacks on Hybrid Clouds by the distributed denial of service. The authors first published an investigation into cybersecurity problems and the effect on cloud environments. The authors analyzed the study of cloud infrastructure, denial-of-service, and malware identification and mitigation methods released between January 2010 and December 2020. Current strategies to prevent distributed service denial attacks were tested by the authors. The authors then developed and introduced a stable framework for networks that mitigates distributed service denial attacks on hybrid cloud environments. The proposed infrastructures and the findings contrasted with the single data center architectural architecture was carried out at the network and device level assaults. This chapter further discusses future study directions.

Chapter 8: Classifying Cyberattacks Amid Covid-19 Using Support Vector Machine

Internet plays a dominant role amid the Covid-19 pandemic as to meet day-to-day activities as education system, financial transactions, businesses, and social gatherings started to function in online mode, leading to tremendous use of networks peaked to the level of cyberattacks. Simultaneously, the thirst for finding the data related to Covid-19 in order to take necessary precautions gave rise to huge risk of cyberattacks by browsing Covid-19-related websites, apps, and falling into the trap of attackers risking the systems security. This research work considers the tweets related to cyberattacks and classifies using machine learning techniques and analyzes the impact of this pandemic. It was observed that Support Vector Machine yielded high accuracy of 94% in classifying Covid-19, followed by decision tree with an accuracy of 88% among other classifiers. The results were evaluated on different metrics like error rate, precision recall, and F-Score, and SVM yielded high results among all.

Chapter 9: Cybersecurity Incident Response Against Advanced Persistent Threats (APTs)

Recent technological innovations and new age computing models in IT infrastructure have provided faster bandwidth speeds, cloud computing, mobile computing, and virtualization, which have virtually melted the boundaries between traditional on premise and internet-based enterprise security perimeter. This has created a data-rich digital era, which is in fact an excellent opportunity for hackers and threat vectors leading to cybercrime. Advanced Persistent Threat (APT) is a highly sophisticated threat. Initially, such attacks focused and targeted government, state, or financial institutions only. However, recent breach reports and studies have started to indicate the trend of APT involving wider domains. This chapter takes a critical look at the impact and incidents due to APT and the advanced evasion techniques for packing, encryption, and behavior obfuscation during APT attacks to hide their malicious behavior and evade detection.

Chapter 10: IoT Architecture Vulnerabilities and Security Measures

Human beings as the most intelligent and skillful creature never end their quench to find easy, smart, and efficient solutions to emerging problems with the growing technology. IOT is one such application which integrates many other technologies within it to deduce the smart and intelligent mechanisms to perform tasks in varying arenas, whether it is day-to-day activities or in problem-oriented activities such as health diagnosis, natural calamities, military applications, education, research, transportation, inventory, agriculture, energy harvesting, forestry, communication, and entertainment. Even though smart and intelligent devices, but the major majorly handle these, all concern that surfaces out is the security and maintenance of these devices, which are susceptible to the malicious networks. This chapter focuses on all the aspects of security concerns related to IOT environment and its devices.

Chapter 11: Authentication Attacks

Internet is known as an amazing platform that changes the manner in which we impart and perform business exchanges in the current technology era. It has now become part of our lives in addition to the fresh security threats it brings forth, prepared to set out towards the journey of destructions. Transmitted information level is turning out to be progressively significant particularly as associations that used to just be completed offline, for example, bank and business trades are presently being done online as Internet banking and electronic business trades, and harms because of such assaults will be more prominent . As expanding measures of individual information are surfacing on the Web, it is basic to stay careful about the dangers encompassing the ease with which our private details can be accessed and exploited. This is where terms like authentication come into picture.

Acknowledgments

I would like to especially acknowledge the editors (Dr. Akashdeep Bhardwaj and Varun Sapra) for all the hard work they put in managing the chapters of the book. Without their collaborative effort, this book would not have gone to publication. We are proud to present the book on *Security Incidents and Response Against Cyber Attacks*. We would like to thank all the reviewers that peer reviewed all the chapters in this book. We also would like to thank the admin and editorial support staff of EAI Publishers that have ably supported us in getting this issue to press and publication. Finally, we would like to humbly thank all the authors that submitted their chapters to this book. Without your submission, your tireless efforts and contribution, we would not have this book.

The proliferation of mobile devices and ubiquitous Internet connectivity has also increased the threat surface area. The digital transformation of our lives, our work, and our social activities is likely to result in more incidents and attacks. Therefore, preparing, planning, and mitigating security incidents and responses to the cyberattacks must become part of an organization and every individuals' daily routines in the "new normal."

I hope everyone will enjoy reading the chapters in this book. I hope it will inspire and encourage readers to start their own research on security incidents and cyberattacks. Once again, I congratulate everyone involved in the writing, review, editorial, and publication of this book.

British University Vietnam   Sam Goundar
Hanoi, Vietnam

References

1. K. Bissell, R. Lasalle, P. Cin, Ninth Annual Cost of Cybercrime Study Research Report. Accenture—Insights, Security (2019), https://www.accenture.com/us-en/insights/security/cost-cybercrime-study
2. S. Morgan, Cybercrimes Damages $6 Trillion by 2021. Cyber Security Ventures, Cyber Security Magazine (2017), https://cybersecurityventures.com/hackerpocalypse-cybercrime-report-2016/
3. M. Pokladnik, *An Incident Handling Process for Small and Medium Businesses.* (SANS Institute Information Security Reading Room, 2020)

4. L. Rosencrance, 10 Types of Security Incidents and How to Handle Them. Tech Target Network—Security Search (2019), https://searchsecurity.techtarget.com/feature/10-types-of-security-incidents-and-how-to-handle-them
5. M. Rouse, Computer Emergency Response Team (CERT). Tech Target Network—Tech Accelerator, WhatIs.Com (2019), https://whatis.techtarget.com/definition/CERT-Computer-Emergency-Readiness-Team

Preface

The title of this book, *Security Incidents & Response Against Cyber Attacks*, is a brainchild of three people—two at the University of Petroleum and Energy Studies, Dehradun, India, and one at British University Vietnam, Hanoi, Vietnam, wanting to make the world a better, secure place, hoping to share ideas in the form of a book. Big thanks to all our coauthors, who as experts in their own domains, for sharing their experience and knowledge. This book is an attempt to compile their ideas in the form of chapters and share with the world. This book provides use case scenarios of machine learning, artificial intelligence, and real-time new age domain to supplement cybersecurity operations and proactively predict attacks that can preempt cyber incidents. In order to prepare and respond, cybersecurity incident planning is highly essential. This starts from a draft response plan, assigning responsibilities, use of external experts, equip organization teams to address incidents, prepare communication strategy, and cyber insurance. Incident plan involves classification and methods to detect the cybersecurity incidents and includes incident response team, situational awareness, and containing and eradicating incidents right until cleanup and recovery.

Dehradun, India
Dehradun, India

Akashdeep Bhardwaj
Varun Sapra

Acknowledgments

I would like to start by thanking first my parents and then my wife for their immense support and giving me the time and space to work on this book and my research. My wife looked after the house, groceries, and kids to reading early drafts and giving me advice, which was positively useful. A big shout-out to the European Alliance for Innovation (EAI) team and especially my Managing Editor, who kept up with my constant spam emails and diligently guided and supported me. I am honored to be a part of this book journey with all my coauthors, so thank you for letting me serve, for being a part of EAI, and for putting up each time with my irritating endless emails and help; the authors turn their ideas into stories.

About the Book

This book provides use case scenarios of machine learning, artificial intelligence, and real-time new age domain to supplement cybersecurity operations and proactively predict attacks that can preempt cyber incidents. In order to prepare and respond, cybersecurity incident planning is highly essential. This starts from a draft response plan, assigning responsibilities, use of external experts, equip organization teams to address incidents, prepare communication strategy, and cyber insurance. Incident plan involves classification and methods to detect the cybersecurity incidents and includes incident response team, situational awareness, and containing and eradicating incidents right until cleanup and recovery.

Advanced level of cybersecurity attacks are frequent, targeted, well researched, fully resourced, financially motivated, and usually lie undetected and persistent. While data and system owners including IT professionals from Security Operations, Network, and Firewall teams or top management executives are bound to get a heart attack on hearing the keywords (like Server Hacked, Cyberattack Incident, Malware inside systems) or seeing similar emails flying around, there are ways to regain sanity. Most organizations gather and have at their disposal a log of machine generated logs and data.

The book shares real-world experiences and knowledge about how to go about during a breach or attack. The book describes security attacks, trends, and scenarios along with attack vectors for various domains and industry sectors. Some of the concepts covered in the book are New Age Cybersecurity Attacks, Malware and Phishing, Cyber Physical Attacks, Cyber Safety Techniques, Network Security, Industrial Security, Blockchain Security, and Incident Response.

Contents

Chapter 1
By Failing to Prepare, You Are Preparing to Fail

Dinesh O. Bareja

1.1 Introduction

It is common to see organizations worried about cyber security and their continuous effort to improve their risk posture. Practically every security standard, best practice, regulatory requirement, or industry guidance asks for robustness, resilience, and continuous improvement in order to report compliance. One of the prescribed controls is Incident Management, which is also included when complying with the requirements, however therein (usually) lies a catch. Organizations tend to go easy in setting up their Incident Management (IM), or Response (IR), functions. The reason is easily identified—security incidents do not happen every day, especially serious ones where the disruption requires the might of the organization for recovery. The cause for worry is justified, considering the continued sprawl and complex evolution of technology in the enterprise. A common cliché in security circles goes "There are only two types of companies—those that know they've been compromised, and those that don't know. If you have anything that may be valuable to a competitor, you will be targeted and almost certainly compromised."

It is a known fact that a breach, or a security incident, can result in a disruption of business activity. Any major (or minor) disruption is not welcome as this is a cause for unforeseen expenses, loss of reputation and clients, regulatory penalties, or legal action by stakeholders. For example, Equifax announced a breach in their systems in September 2017 and are still grappling with the settlement costs which

D. O. Bareja (✉)
Founder & COO - Open Security Alliance, Bombay, Maharashtra, India
e-mail: Dinesh@Opensecurityalliance.org

© The Author(s), under exclusive license to Springer Nature Switzerland AG 2021
A. Bhardwaj, V. Sapra (eds.), *Security Incidents & Response Against Cyber Attacks*,
EAI/Springer Innovations in Communication and Computing,
https://doi.org/10.1007/978-3-030-69174-5_1

had touched about $900 million in 2019, and seem to have spent $1.4 billion (Equifax Data Breach FAQ n.d.) in upgrading their security. The root cause for the breach was the compromise of a known vulnerability in Apache, which had not been patched by the Equifax team. Another worrisome issue in security breaches is the "dwell time" of the hacker, in the network, before the breach is noticed, or identified. The industry average is 197 days and it is anyone's guess what damage can be done by a malicious actor who is hidden in the network during this period.

Enterprises should plan for resilience to be built into their security processes/ posture to ensure survival in the event of a breach. As such, Incident Management, Business Continuity Management (BCM), Disaster Recovery (DR), Crisis Management must form part of the enterprise Information Security Management System (ISMS) and will have to be invoked in the event of a security incident, or disruption. Each of these functions has a crucial role to play in ensuring the resilient recovery and continuity of business, but they depend on effective incident response and management. Organizations should have an effective Information Security (or Cyber Security) Management System (ISMS) in place, preferably aligned and compliant with global guidelines, or standards (Cyber Security Standards n.d.), like ISO27001, NIST Cyber Security Framework V 1.1, COBIT or frameworks like PCI-DSS as illustrated in Fig. 1.1 below.

Mature ISMS will incorporate an effective IM/IR function as a critical function and take steps to keep it well oiled. IR is a skill, multi-faceted skill that must be seriously nurtured, planned, and prepared for. Incident Management is the discipline for planning, preparing, identifying, responding, recovering, analyzing, and learning from security incident(s). A typical IM process is illustrated in the accompanying Fig. 1.2 below.

Fig. 1.1 NIST Cyber Security Framework V 1.1, COBIT

Fig. 1.2 Incident Management Process

Every component in this function requires multiple skills. In the normal course of setting up the Cyber Security, function, and associated governance structure in the organization, one will usually see the IM component as an adjunct to the Information/Cyber Security office. The Incident Manager may be filled as an independent role or any one person in the team may assume additional responsibility, which depends on the size of the organization, perceived risk, and volume of events/incidents handled. Whatever be the case, the importance of planning and preparation cannot be discounted, as with any task. The IM function can only be effective in the face of an attack, disruption, or any security incident only if the organization has planned for various eventualities and is prepared to respond swiftly and surely. This requires diligent attention from the initial stages of risk and threat assessment where mitigation actions are identified. This translates into the response strategy for different scenarios and for an organization will require different playbooks for each department.

Incident Response is a function within the IM organization and maybe internal or outsourced depending on the skill and resource requirement. A security incident response may call for containment, recovery or for investigation and any of these activities may require skills in forensics, networking, log analysis, malware analysis, etc. which may or may not be available internally in the enterprise. As such, IR needs to be planned and tested and this again is another important task to take up in the planning and preparation phase, as we will see further in this chapter. Planning and preparing for an effective incident response starts with the requirement to have a well-defined policy, a comprehensive set of procedures, and an aware governance structure to ensure the operational success. These must be aligned to the existing procedures and practices prevalent in the organization without trying to induct new systems, techniques, or activities. The most effective plan is one, which is practical, pragmatic and seeks to leverage the day-to-day workflow and replicate the mode of carrying out the daily tasks. The National Institute of Standards and Technology (NIST) publication NIST 800-61rev2 (Computer Security Incident Handling Guide n.d.) provide a definitive guide for IM. The workflow in a typical IR, as per NIST guidance, will be to take the four steps as illustrated in Fig. 1.3 below, but as has been said earlier the "preparation" phase is usually given short shrift saying this is more a compliance issue and this results in weakness cascading through the remaining lifecycle.

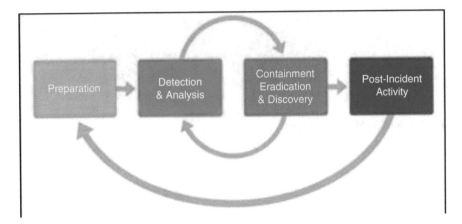

Fig. 1.3 NIST Incident Management Steps

1.2 Plan & Prepare

Cyber security is, in a manner, fighting a continuous battle against unknown malicious perpetrators and is a big drain of resource for any organization. Regular business days can be termed as peacetime and the hyperactive period of detecting and responding to an incident as wartime. In order to be able to put up a strong defense, or an effective response, *organizations have to prepare for war to stay at peace*!

Every action in the IM/IR lifecycle needs to be well planned and prepared for. This will ensure an efficient recovery from an incident. This "peacetime" activity should be used to carry out important prep activities such as identifying needed skills, training the IM/IR team, getting commitments from outsourced stakeholders, and obtaining internal management support. Another part of the planning and preparation is the conduct of drills and exercises as this will help keeping teams on high alert, and provide the opportunity to practice and hone their skills. The fundamental elements in planning and preparation cover the full range of the Information/Cyber Security practice closely meshed with business requirements illustrated below in Fig. 1.4.

1.3 Strategy & Governance

Well planned is half done! The first activity is to envisage and design the structure of the IM/IR unit in the organization. Will it be part of the IT Committee, the IS Committee or will it be a separate unit will depend on organization business, size, and risk. However, an Incident Manager must be identified who will be the leader for the IR activity.

Figure: Cybersecurity Incident Planning and Preparation Activities

Fig. 1.4 Cyber Security and Business Processes

The IM team is like a crack commando unit, which will be called upon to respond to incident(s). This team must have all round skills, well coordinated and ready round-the-clock. This does not mean that the organization has to hire a full team but can identify team members from within the organization. The team should be multi-disciplinary, so it will comprise members from all departments, including HR, Administration, and Finance, though they may ask why. Planning and preparation activity should also include external parties like Managed Service Providers (MSP), Managed Security Service Providers (MSSP) among others. An IM/IR strategy should be put together to set the direction for creating and operationalizing the IM organization. The IS, IT, Risk leaders, along with senior management have to decide whether to outsource the IM/IR function to an MSSP. Outsourcing can be in part or in full and some organizations may only contract for specific services like IR or Forensics.

The output of this activity will be the formation of an Incident Management Team/Committee, which will have the skill and resources to respond to any type of disruption to the business operations. A typical team composition and structure is illustrated in the table below, and should be tailored for the organization as presented in Table 1.1 below.

While it is important to have business leaders as part of the IM team, it may not be practical for the leaders to be involved in the operational activities. Hence, members with domain skills and knowledge are inducted from the business and functional teams, and buy-in/commitment is sought from the business/functional head. While the business/functional heads may not be a full-time member of the IM organization, it is necessary for them to attend IR training and awareness sessions, and participate in drills, to understand the IM/IR lifecycle, and to provide their inputs. Like any organization, IM/IR will have members who are responsible for operations and those who have oversight on the strategy, effectiveness, reporting, and compliance and business issues. The categorization illustrated in the table is representative of the role and responsibility structure, which should form the IM/IR team.

Table 1.1 IM/IR Team Stakeholders and Composition

Incident Manager

(This position is usually held by the Information Security Manager/HoD or a senior resource who is hired by the organization for the role and has the skill and expertise for Incident Management and Response)

Department representation	Unit	Interested parties	Unit	Stakeholder representation	Site	Government stakeholders
IT Manager/HoD		CEO		Vendors		CERT
IS Team Members		Board Members		ISP		Critical Infrastructure
BC/DR Manager		Banks		Alternate provider	Site	Cyber police
Business members	Unit	All Business Heads	Unit	Facility provider		Fire Station
Finance		Customers		MSSP		Utilities
HR		Training		MSP—Forensics		Regulatory Authorities
Risk		All Employees		MSP—IT support		National Cybersecurity Advisor
Legal and Compliance		All Contract Personnel				
Insurance		Vendors				
Crisis Manager						
Sales						
Account Management						
Purchase						

1.4 Departmental Representation

This represents the various departments in the organization that must nominate one or more persons to the IM/IR team. This is necessary to ensure that there is

1. Cross-functional knowledge and skill available at the table during an emergency
2. Person familiar with the department, its working and people that will allow him/her to reach out seamlessly in times of need.

However, all members of the Security team should be a part as well as the Head of IT. The IT and IS department head will be the only full-time members of the IM/IR organization. They should be the full-time members of the IM/IR organization and will be responsible for the security incident operational readiness, preparedness, response, and recovery.

1.5   Interested Parties

These should be informed, periodically, about the activities being undertaken by the IM/IR team towards effective readiness, i.e., how is the enterprise prepared to respond to attacks. These members will be the senior leaders from within the enterprise as well as others, as indicated. Regular communication and information about IM/IR preparedness will provide assurance to these stakeholders and keep them informed about any expectations from them during an incident. While internal parties like CEO, Board, BU heads, and FT employees may be informed about the preparedness in more detail, the message to others has to be customized so as not to divulge internal confidential information. Customers are particularly important, as they must have continuous assurance of continuity of business and services being provided. As an interested party, periodic communication to customers should be planned informing them about tools, trainings, drills, and outcomes (particularly with reference to their business). It may also be a good move to include large customers in IM/IR exercises and trainings. Other groups like vendors, non-FT employees/contract personnel, and banks will be interested to know the depth of resilience in the enterprise. Their business depends on you, and they need to be assured that your business operations will be standing in the face of any disruption. Any business operation needs vendors as much as it needs a customer. Along with these, every department has its place of importance—whether business oriented, functional, and operational or support oriented. To overlook any is a grave error.

1.6   Stakeholder Representation

Inclusion of external stakeholders in the IM/IR team is important but should be selective. Since IM/IR activity (at all, or, at any time) will provide exposure and discussion on business recovery, it is a given that internal confidential information will be tabled. In view of this occurrence, it is necessary to evaluate the risk of including stakeholders in the IM/IR team, especially as full-time members/participants. At times, vendors or external resources may be embedded within the organization and working along with FT employees. While one may think that it is okay to include the vendor/external resource in the IM/IR team, as a full-time member, it will be advisable to revisit their contractual agreements and amend the same appropriately to include the new responsibility. Additionally, Non-Disclosure Agreements (NDA) may need to be updated. Organizations may have outsourced technology or security operations, in full or part, and these stakeholders should also be included in the IM/IR team. Depending on the level of their involvement in business, a decision should be taken. In case of MSSP, it is necessary to have their representative in every meeting, however, discretion is advised when discussing any internal confidential information like finance, business plans, etc. Parties like ISP, alternate site providers, and facility providers should be invited from time to time, to join the discussion and to participate in drills and tests. While their presence may not be necessary in every meeting or as a full-time member, their services are essential to the overall IM/IR practice and their business health and preparedness must always be under watch.

1.7   Government Stakeholders

Numerous government entities have been set up as regulatory authorities or for cyber security support and law enforcement agencies that require organizations to report incidents and attacks. Domestic and international laws and regulations require organizations to have reasonable information security practices enabled to protect their digital assets. There may be mandatory requirements for review, audit, disclosure, and reporting with one or more entities. Incident related reporting, legal and regulatory compliance obligations should be included in the plans at the preparatory stage itself. Users, responders, and team members should be aware of the requirements and appropriate responsibilities should be assigned to identified members in the team, with inputs from the legal department members. The premium on security skills is high and it may be an expensive decision to hire skilled resources. Besides, organizations may not need all the skills on a regular basis. For example, digital forensic expert may be required only at the time of investigation of an incident or for data recovery from damaged system, and such instances are few. As such, IM/IR functions may be outsourced, as it will help the organization to have multi-faceted technology response and security investigation skills on call without the need to invest in hiring and maintaining a big team. The structure of the IM/IR management organization will be in a manner to have representation from all enterprise functions. Roles and responsibilities have to be assigned and appropriate training should be imparted to the team members.

Typical roles or areas of expertise will be:

– Networking traffic analysis
– Infrastructure devices (hardware)
– Application security
– Malware analysis
– Threat hunting and intelligence
– Investigation techniques
– Digital forensics
– Log analysis

The Incident Manager has to assign roles to the team members as per their areas of expertise. In event of non-availability of the skill within the organization, one should consider outsourcing or entering into a contract with experts who can join the team when needed. When entering into a contract to hire when needed, the Manager should assess the skill of the expert to be hired and ensure the availability for training, awareness, and drill sessions. The requirement of special skills should be assessed and manpower resources should be provisioned accordingly - this can be achieved by training internal team members, or hiring subject matter experts on contract.

1.8   Develop Policy and Procedures

Once the strategy has been decided and the IM/IR governance established, it is time to document the IM/IR policies and procedures. Policies, procedures and forms, once documented should be reviewed annually at least and will refer to IS policies that are being followed in the organization. Any organization with an operational Information/Cyber Security function will have, at the very least, an Information/ Cyber Security Policy, Risk Assessment Framework and Register, Asset Register, Access Register among others. The documents that should be developed for the IM/ IR function should include the following:

- Incident Management Policy
- IM Committee Terms of Reference
- Root Cause Analysis Template

> *The policy document should follow a convention wherein statements containing the words "shall" and "required to" in the document are mandatory rules. Failure to observe these rules may be construed as non-compliance to the policy. The statements containing the words "should" and "recommended" imply a desirable requirement. Failure to adhere to these rules may not be a direct non-compliance.*

The objective of the IM Policy is to provide a framework for management of information security incidents, providing guidance to:

- Detect/identify security incidents proactively
- Respond to security incidents efficiently and effectively
- Contain the damage due to incidents
- Crisis management and communication
- Invoking the BCP/DR plans
- Work with technology team to bring systems back to a normal state
- Follow up actions to investigate incidents and root cause analysis (RCA)
- Identify weak control structures, and if necessary, update
- Prevent recurrence and minimize damage by enabling improvements based on RCA
- Facilitate constant improvement cycle of the ISMS
- Post incident management and compliance reporting

While writing the IM Policy it is important to include, at least, the following sections: *(please note that the sections are not in any particular order and are presented below randomly)*

- Introduction—general introduction for IM and IR and the purpose and coverage of the document. Include a statement on the importance of IM as well as management support for information/cyber security.
- Statement of Purpose—a management statement on the purpose of the policy.
- Scope—covering the locations.
- Policy Statements—address the following ISO27001 specifics:

– ISO27001:2013 Clause: Information security incident management (Clause 16.1 Management of information security incidents and improvements). The following controls importantly:

> 16.1.1 Responsibilities and procedures—Provide details of the IM/IR team, with roles and responsibilities that have been defined.
> 16.1.2 Reporting information security events—Process for reporting and security incidents and escalation based on the classification.
> 16.1.3 Reporting information security weaknesses—Process for reporting security incidents by employees, etc. This is based on awareness and training of employees which should be referred to.
> 16.1.4 Assessment of and decision on information security events—Procedure for assessment of incidents and events in the SOC or in an outsourced environment.
> 16.1.5 Response to information security incidents—The workflow and process for incident response. This should be based on the classification and the scenarios.
> 16.1.6 Learning from information security incidents—Procedure for post incident analysis (root cause analysis) and updating controls based on the learning from the same.
> 16.1.7 Collection of evidence—How to collect evidence maintaining the integrity of the same.

The above will help to attain compliance with the requirement of the standard, specifically; however, there are yet a number of areas, which should be included in a policy. It should be as detailed as possible to address the size and volume of threats for which the organization may be at risk.

As such, the following should also be addressed in the policy:

– Training & Awareness

> Training for IM/IR committee members.
> Security incident awareness for all users.
> Training for response team members.
> Reference may be made to the requirement for the IM/IR team to undergo BCP/DR training too.

– Classification Guidelines

> The policy should provide guidance on classification like critical/high/medium/low/info, and specify what each level implies. Response time should be defined for the classification and responsibility assigned where possible.

– IM Lifecycle

> Provide a systematic breakdown of the lifecycle and the activity to be carried out at each step.
> NIST Cyber Security Framework provides a comprehensive lifecycle definition through the following steps: Identify > Protect > Detect > Respond > Recover.

In order to go granular, one may use a more detailed lifecycle by identifying major activities under the NIST guides. One suggestion will be: Identify > Classify > Evaluate > Contain > Eradicate > Restore > Review > Learn and Update.

– Incident Investigation

> Procedure for forensic investigation with responsibility assigned.
> If this is an outsourced activity, the SLAs and expectations should be included or referenced.

While creating policies and procedures, one should take care that these documents are not too lengthy else, reader attention is difficult. The solutions to make specific "playbooks" which provide the policy and procedure for particular functions and the collective of playbooks make up the complete policy/procedure set. This will make it simpler to use in operations or in training.

1.9 Team & Resourcing

The IM/IR capability in the organization can be considered ready to meet any eventuality if it is staffed with professionals and experts having adequate skills across the security domain. The policy document should also have the organization structure of the IM/IR team and a leader should be designated. This has been addressed in the requirement for having a proper governance structure.

An effective IM/IR organization requires the concerted involvement of various teams in the overall incident lifecycle. The team will typically consist of members from senior management, support functions as well as general employees who should promptly handle an incident so that containment, investigation, and recovery can be done quickly.

The following roles must be filled with properly qualified personnel:

- Incident manager
- Help desk personnel

  – Identify nature of incidents based upon defined classification and categorization.
  – Prioritize incidents as per impact and SLA guidelines.
  – Responsible for incident closure at level 1.
  – Delegate responsibility or escalate to next level in event of inability to close.

  > Delegates responsibility by assigning incidents to the appropriate internal group.
  > Escalates incident to next level (L2, L3) as per the response workflow.

  – Obtain closure reports and update the helpdesk.
  – Prepare reports showing statistics of incidents resolved/unresolved.
  – Perform post-resolution review to ensure that all work services are functioning properly and all incident documentation is complete.

- Incident management team comprises technical and functional staff involved in supporting services, for carrying out the following:
  - Correct the issue or provide a work around.
  - Ensure the incident is resolved within the defined SLA and systems are recovered.
  - Provide functionality that approximates normal service as closely as possible.
  - If this is a reoccurrence, escalate to problem management team.
  - Carry out a forensic investigation to ascertain the attack.
  - Update the threat library to identify any similar attacks in future.
  - Conduct a root cause analysis to identify control weaknesses in the IT infrastructure.
  - Provide the reports and update processes to avoid recurrence.

1.10 IM/IR Skill Requirement

Incident Management and Response requires multi-faceted skills in the team, ranging from management to technology to forensics, data analytics, etc. As such, in planning and preparing for setting up the Cyber security IM/IR capability one must ensure that such cross-functional skills are available. Organizations may not have all the skills available in-house and can outsource or obtain on contract or on call. Outsourcing of skills means entering into an agreement with an individual or entity, which can provide the necessary expertise when called upon. They may make the estimate for the response at the time of being called and the Incident Manager will need to make immediate negotiations. On contract means that the IM/IR team contracts the professional services on retainer terms and the person/entity is bound to provide a defined number of hours, or respond to incidents. On call is a loose, or informal, arrangement where the expert is called when needed for the professional services. In order to take advantage of this arrangement, make a list of available experts and network with them to make relations and map their skills. It will be advisable to ask them if it is okay for them to join on short notice and then make your call plan accordingly.

1.11 Outsource Vendors

Outsourcing is a function, which is finding high levels of acceptability across all business sectors due to the value, and benefits that accrue. Organizations may use outsource service providers for numerous reasons, including, but not limited to

- Cost Benefits: Instead of going through the process of hiring and then managing and retaining highly skilled employees, it is easy to have a provider who will provide the mix of skills required and will be responsible for the numbers.

- CAPEX to OPEX: One of the biggest benefits is that the organization can avoid capital expenditure for acquisition of hardware/software to set up a SOC or the IM/IR capability.
- Transfer of Responsibility: In an outsourcing scenario the organization transfers the responsibility for security enablement, monitoring, and response to a third party, which has the skill and resources for the same.

Outsourcing the IM/IR function means engaging with a third party that has the experience, skill, and resources to manage and respond to different incident scenarios, which may manifest themselves in the organization. In order to finalize an outsource vendor, an in-depth assessment of their capability in terms of the people, process, and technology should be carried out. The critical component in the outsourced engagement is the people because the incident response/forensic experts with real skills are available at premium costs. In addition, one goes into the outsourcing mode to first offset this challenge. The experts on the outsource vendor team should be seasoned Digital Forensic Incident Response (DFIR) professionals and the cross-functional expertise should include (at least) digital forensics, investigation, OSINT, dark web, VAPT, application security testing, log analysis, threat hunting, etc. It must also be noted that one does not become an IM/IR expert by completing a certification or two and that real-world experience is necessary to garner skills to be effective in response, containment, and recovery. The good experts in the outsourced vendor team will have been through multiple response and recovery situations in different verticals. They should have industry standard certifications and be able to demonstrate their expertise in their body of work.

However, engaging a vendor one should also evaluate the cost against the perceived risk. The skills being included in the service package may be adding to the cost of the outsourcing service and the buyer is advised to ask for a cost breakdown before signing the agreement. Other factors to look out for are:

- SLA provided by the vendor (*this has to be matched with the internal SLA and the requirements of own internal and external customers*)
- Availability (*this should be 24*7*365 or at least cover one shift especially the one where you need to have eyes on your infrastructure*)
- Warning/Alert systems (*how and when does the vendor alert you and other team members in the event of an incident*)
- Solutions and technologies (*ticketing, threat and vulnerability management, analytics, SIEM, DLP, etc.*)

1.12   IM/IR Training & Awareness

Security awareness is a critical need in any resilient and effective information security setup in an organization. It means that employees, stakeholders, management, etc., all should be aware about security risks and threats and be able to identify common attacks, or anomalies in the workplace. This is also true about personnel in specialized roles that they need to be trained and aware of their role, responsibility and have practiced skills, even if they have the experience. As such, at this early stage of planning and preparation to set up the IM/IR practice and organization, it is necessary to create a training and awareness program that is focused on imparting IM/IR and domain specific knowledge to the team personnel. This plan should include periodic sessions on adhering to the IM/IR policies and SOP, along with the touch points with the BCP/DR plan. Training and awareness program should be detailed and can be created by

(a) making a list of the sessions and activities using the example below
(b) map the sessions on an annual calendar
(c) identify the in-house and external personnel who must participate
(d) Periodic assessment of the personnel through practice and refresher sessions

The list below may be used as a road map to build the internal course and session development map.

- Training the team (*an SOP with a list of learning resources*)

  – Mentoring
  – Self-study
  – Courses
  – Library
  – Exercises

- Testing the team/procedure (*learning assessments using labs, etc.*)
- External coordination (*procedures and statutory requirements*)

  – Law enforcement
  – Media
  – Other IR teams (data center, ISP, etc.)

- Managing incidents (*IM/IR operations*)

  – Incident response
  – Evidence
  – Assigning incident ownership
  – Prioritization and containment
  – Tracking response

- Containment strategies (examples)

  – Shutting down a system
  – Disconnect from the network
  – Change filtering rules of firewalls
  – Disabling or deleting compromised accounts
  – Increasing monitoring levels
  – Setting traps/deception

- Adhering to containment procedures
- Record all actions
- Reporting
- Root cause analysis

1.13   Incident Drills & Testing

Another activity to plan and prepare for is drills and testing which should be done periodically around the year. The activity helps keep the IM/IR skills and processes in current practice as well as helps identify gaps, including gaps in technology tools that have been implemented. There are different types of activities to carry out and one must include the following in the planning:

- **Security Drills** *(different disruptive scenarios are envisioned, and the IM/IR teams get together to respond and manage).*
- **Red Teaming** *(a known hacking team "attacks" the infrastructure and the internal IT/IS team puts up the defense. The skill is in being able to identify the attack and then to respond/contain/recover—all done in a test environment).*
- **Vulnerability Assessment and Penetration Testing (VAPT)** *(while this is a regular activity in the Information Security calendar it should be included in the IM/IR plan for obtaining the report and assessing weaknesses that have been identified).*
- **Phishing drills** *(this test is also usually included in the information security department plan; however, it is included here as an item which must be part of the IM/IR team to understand and analyze all vectors).*

Drills: A plan for an incident response drill should include activities, which will help bring about visibility on the level of preparedness for the management and response objectives of the IM/IR organization. Typically, the plan should include (in no particular order):

- Liaison and communication with BCP/DR team and invoking the crisis management team.
- Communication channel and methods—who will be contacted during an incident for management and response (phones, contact information, call trees).
- Availability/Provisioning of hardware and software.
- Incident analysis resources (Documentation, network diagrams, etc.).
- Detection and analysis of the incident (scenario) has occurred and prioritization.
- Activity of Containment, Eradication, and Recovery.
- Reporting on the drill and the outcomes to management.
- Analysis of the drill outcomes—documentation of procedures followed, gaps/issues/challenges.
- Learning from the drill along with remediation suggestions.
- Plan for updating controls and processes based on the learning and gaps observed in the drills.

While a full-blown drill, as above, may require allocation of hardware/software and labor resources, which may not be easy to come by in a running business enterprise, the importance of conducting drills periodically cannot be overlooked. One can take an alternate route and conduct modular tests, which address one or more parts of the business enterprise. For example—a test of the web sites and web facing infrastructure, or a test for ransomware attack, etc.

Drills/tests can be in the form of

- **Paper tests, which are mostly theoretical,** can be used to look for small process changes; they can also be used to gauge the skill level of team members.
- **Tabletop Exercises** are just that stakeholders are invited to engage in a security event scenario. While such an exercise may appear informal, it raises the level of awareness of ineffectiveness or ignorance of role/responsibility and post incident chaos. The plan should include the right stakeholders to participate and to present a real-life scenario or as real a threat scenario as possible.
- **Simulated Attacks** will be akin to a red-teaming exercise where conditions are simulated that exist in real life. Usually for training purposes it is the most effective way of pressure testing the IR, processes to see how an organization responds when hit by an external threat.

1.13.1   Red Teaming

Red team assessments are a highly effective IM/IR testing method as it is designed to mimic a real attack. The IM/IR planning should include a red team exercise to be carried out at least once a year, or more.

While including this in the plan and identifying the individual (or entity) that will be undertaking this activity, one should remember that an effective red teamer should be able to launch complex attacks and the home team should be trained enough, and prepared, to identify the attack and then respond. The red team will make their attack attempts using various tactics and techniques (social engineering, phishing, brute force, VA followed by PT, stealth attacks, DOS, etc.) and the skill of the in-house/home team is in being able to identify attacks as well as to respond to them. One will have to plan for resources in a separate test environment to carry out this highly destructive testing. While identifying the resources for red teaming, one must ensure that the red teamer does not land up doing a simple VAPT and pass it off as a red team exercise, as has been observed to happen in many engagements.

1.13.2   VAPT

This form of testing is by far most common and is a part of the enterprise ISMS activity. However, as part of the IM/IR planning it will be prudent to check on this and ensure that the VAPT activity will be comprehensive and will cover the IT infrastructure, applications, web properties, mobile apps and that the IM/IR management committee is also a recipient of the VAPT findings and action taken reports. IM/IR team meetings should include an analysis of the findings from VAPT activity and chart possible internal threats and risks. These findings will also help to build scenarios for tabletop testing or drills to assess the effective remediation of the vulnerabilities discovered.

1.14   Conclusion

Being well prepared means, half the battle against any form of attack is won!

Preparing and planning the IM/IR function requires a high level of diligence in order to create an effective response system, which can take care of unwanted incident situations. Any incident, small or big, is a drain on the business, and should be mitigated to recovery in the shortest possible time. This is where an effective response system comes in, as it will facilitate a quick recovery, ensure that this is resilient, and avoid reoccurrence. Every component discussed needs to be addressed and has its own place in the larger scheme of the IM/IR. Building the strategy and establishing a proper governance structure will create the foundation for an efficient IM/IR capability, whether it is in-house or outsourced. Most organizations do exert the effort but not the diligent effort to create the in-house entity and then things go wrong in the unfortunate event of an incident. Besides, the IM/IR meetings stop discussing strategy and new threats and start focusing on the operational issues, with reducing attendance.

Yes, this is another challenge, which has to be addressed at this early stage—the buy-in from all department heads, members of IM/IR committee and such stakeholders in respect of the criticality of the role being assigned. ToR for committees and meetings must include mention of compulsory attendance to meetings, training, and awareness, else the essence of preparedness is lost and there will be chaos in the event of an incident.

Following up on a good strategy, one should have strict policies, well-defined procedures and playbooks to set the direction. With a good mix of resources, tools, and tactics, most attacks can be foiled or contained. In this respect, training and awareness should be a regular activity for the team members (both internal and external). It should be noted that IM/IR is not a stand-alone practice and must continuously coordinate and course-correct with the other processes and practices in the enterprise IT, ISMS, and BCP/DR areas. It is only with the effective coordination among all these departments and, of course, the business and functional units that the battle against security incidents can be won.

Organizations need to have adequate and appropriate tools and technologies to identify security incidents as they manifest on the network, in any manner (malware, brute force, stealth) and then the resource and skill to respond and contain them. Effective planning and preparation for cyber security incidents, with consideration of all touch points in the IT/IS environment as well as with business and functional units will be a strong unit in ensuring quick recovery and success for ensuring continuity of business.

References

Computer Security Incident Handling Guide. https://nvlpubs.nist.gov/nistpubs/SpecialPublications/NIST.SP.800-61r2.pdf

Cyber Security Standards. https://en.wikipedia.org/wiki/Cyber_security_standards

Equifax Data Breach FAQ: What happened, who was affected, what was the impact? (n.d.), https://www.csoonline.com/article/3444488/equifax-data-breach-faq-what-happened-who-was-affected-what-was-the-impact.html

Chapter 2
Design of Block-Chain Polynomial Digests for Secure Message Authentication

P. Karthik, P. Shanthi Bala, and R. Sunitha

2.1 Introduction

The internet has transformed our life into a further dimension, and the arrival of the cloud has completely changed the conventional data storage and access mechanisms. In modern computing, most of the data are stored in remote servers, and they are not under the control of the owner. Therefore, ensuring the integrity of the remote data is crucial for the protection of data. The hash or digest functions come in aid to comprehensively address this issue. The digest function is designed with a one-way property. It is directly unintended for protecting the data but it helps the user to know the integrity violations of the remote data. It processes an input string of any arbitrary length and produces a constant output (Preneel 1998). The output is produced as hexadecimal digits and is typically viewed as a compact form of the given input string. Therefore, the hash function could be considered as a somewhat densification function. In general, the input to output mapping of the digest function is expected to be one-to-one to uniquely distinguish one input string from the other. But, achieving one-to-one mapping on a compression function remains an infeasible task. This is because the compressive nature would force the hash function to produce hash collisions when it is attempted for an attack called brute force. However, applying brute force on a hash function stays a computationally infeasible task when it demands a minimum of 2^{56} iterations (Merkle 1989). Therefore, the cryptographic digest functions try overcoming the threat of brute-force and birthday attacks by the suitable selection of the output size. The modern cryptographic digest functions would choose the minimum output size as 256-bits to exhibit formidable resistance to brute-force and birthday attacks. Also, the digest functions try

P. Karthik (✉) · P. S. Bala · R. Sunitha
Department of Computer Science, School of Engineering and Technology, Pondicherry University, Puducherry, India

© The Author(s), under exclusive license to Springer Nature Switzerland AG 2021
A. Bhardwaj, V. Sapra (eds.), *Security Incidents & Response Against Cyber Attacks*,
EAI/Springer Innovations in Communication and Computing,
https://doi.org/10.1007/978-3-030-69174-5_2

incorporating a stochastic behavior while producing a digest value to survive from a differential attack. This property would enable the digest function to produce bizarre output in response to an insignificant change happening at the input string (Kam and Davida 1979). A. F. Webster recommended strict avalanche criteria for the hash function to survive from differential attack (Webster & Tavares 1985). All the modern digest functions try infusing the aforesaid features in their design to exhibit a strong immunity to the collision, pre-image collision, and second pre-image collision.

2.2 Background

Ralph C. Merkle and Ivan Bjerre Damgard had independently proposed a mathematical model for the block iterated digest function using the keyless and keyed approach, respectively. Their design principles were collectively called MD construction (Damgård 1987; Merkle 1989). The modern cryptographic digest functions like MD5, SHA-160, SHA2-224/256, SHA2-256, SHA2-224/512, SHA2-256/512, SHA2-384/512, and SHA2-512 use MD principles to produce a message digest. It was during 2012; a contemporary standard was introduced for the digest function called SHA-3 (Bertoni et al. 2013). The SHA-3 differs from its predecessors in the construction principle. It uses sponge principles to produce a message digest (Bertoni et al. 2009). Ueli Maurer had proposed an in-differentiability framework to measure the structural weakness of the digest function (Maurer et al. 2004). J. S. Coren had advocated that the claim of no structural flaw in the MD construction might be incorrect for the iterated hash function (Coron et al. 2005). Thomas Ristenpart had opined the in-differentiability and Random Oracle are different. He proved a secure model in Random Oracle could be busted in the indistinguishable framework (Ristenpart et al. 2011). The work of Thomas Ristenpart appeared to continue added evidence for the claim made by J. S. Coren. Saif Al-Kuwari suggested some key security attributes for a hash function to be considered for cryptographic use. They represent collision resistance, pre-image resistance, and second pre-image resistance (Al-Kuwari et al. 2011). The application of elliptic curves and the chaotic functions would continue as the other alternatives for the design of digest functions.

   J. S. Teh had put forth a chaotic model implementing the fixed-point representation. His design used keys to produce a message digest. The previous attempts on chaotic hash functions have been performed with floating-point representation. However, those approaches were experienced with interoperability issues. The JS Tech model was based on the single-dimensional map function and Feistel structure (Teh et al. 2019). Hongjun Liu offered a hyper-chaotic digest function by incorporating the Lorenz system. The system was employed to produce a sponge effect to generate 256, 512, and 1024 bits digests by operating the variables (Liu et al. 2019). Yantao Li had suggested a chaotic digest function constructed from the active S-Boxes. The author had employed the linear chaotic map module to derive four initial values.

These values were later employed on the input blocks to produce the intermediate message digests. The final hash value was got from the cascaded value of the intermediate results and the hash value of the ultimate block (Li et al. 2016). N. Diarra had designed a deterministic hashing based on Elligator's model using elliptic curves. They had produced invertible encodings that were much similar to uniformly distributed Random Oracle (Diarra et al. 2017).

2.3 Motivation

The functional analysis of the standard digest functions shows that these functions employ bitwise AND, OR, XOR, MOD, and SHIFT operators to envision erratic behavior in the hash output. Table 2.1 presents the functional analysis of the standard digest functions.

The MD5 and SHA-160 algorithms were broken during 2004 and 2005 (Wang and Hongbo 2005; Wang et al. 2004, 2005). The partial attacks on the SHA-2 algorithm were carried out to a maximum of 52 rounds (Khovratovich et al. 2012; Sanadhya and Sarkar 2008). The third round report submitted to NIST reveals that the new SHA-3 algorithm was broken up to 5 rounds (Chang et al. 2012). The thorough analysis of the attacks/partial attacks on the standard digest functions reveals the following facts.

- The protection of the digest module depends on the digest size.
- The attacks and partial attacks are independent of the type of construction principle deployed.
- The partial attacks on the SHA2 family and SHA3 family were happened because of the extensive deployment of the bitwise operators at their internal structure.

The suggested model addresses this problem through polynomial products. A polynomial function represents an innate unidirectional function and there are no standard solutions available for the polynomial function with a degree greater than

Table 2.1 Functional analysis of standard digest function

S no.	Name of the Digest	Digest size	Total no. of rounds	Operators employed	Name of the construct	No of rounds broken	Status
1.	MD5	128	64	^, I, Rot, ⊕, <<<,⊞	RO	64	Fully broken
2.	SHA-160	160	80	^, I, Rot, ⊕, <<<,⊞	RO	80	Fully broken
3.	SHA2-family	224,256,384,512	32/64	^, I, Not, Rot, ⊕, <<<, ⊞, ⊞	RO	52	Partially broken
4.	SHA3-family	224,256,384,512	24	XOR, MOD, CEIL(x), Log2(x), and Trunc(x)	Sponge	5	Partially broken

three. Therefore, deciphering the polynomial products of degree 64 from the inverse direction remains an arduous task to perform. The presence of the polynomial products at the individual blocks would forbid the cryptanalysts to set up any block-level assaults. The experimental results prove the suggested model exhibits intense resistance to the collision, pre-image collision, and second pre-image collision. Besides, the avalanche response, near-collision response, and the confusion and diffusion analysis prove the model produces a bizarre output even for an insignificant change performed in an input. Therefore, the suggested model could be observed as an ideal replacement for the contemporary digest functions from the perspective of security.

2.4   Analysis of Standard Digest Functions

The digest function produces hexadecimal output. The digest produced appears similar in size, and it is regarded as the constricted form of the given input string. The mapping of input to output is expected to be one-to-one. This part of the chapter attempts to carry out a functional analysis of the standard digest functions like MD-5, SHA-160, SHA2-256, and SHA3-512. This chapter also attempts to identify the vulnerabilities of the standard digest functions from the perspective of security.

2.4.1   Merkle–Damgard Construction

Ralph C. Merkle and Ivan Bjerre Damgard attempted to prove the provable secure properties by employing the block iterated principles with keyless and keyed design paradigms, respectively. Their methods were based on the Random Oracle (RO) model, and it was referred to as MD construction. The RO model would produce a similar output for identical inputs. It would produce a unique output for the other input. The compressive nature of the digest function makes it difficult to achieve the one-to-one property following the pigeon-hole principle. Any digest function that involves compression would generate hash collisions when it is trailed for a brute-force attack. To overcome this problem, the digest function imposes the stochastic attributes in the digest to be produced. In the same way, it would choose suitable digest length to withstand against brute-force and birthday attacks. Merkle opined that any function that entails to the minimum of 2^{56} iterations would be hard to break using the aforesaid methods. This would demand a hash function to contain a minimum of 112 bits in the hash output to survive from brute-force and birthday attacks. The principle of operation for the MD construction is sketched in Fig. 2.1.

The MD design paradigm divides the message M into t equal blocks. These blocks are referred to as $M_1, M_2, M_3,...M_t$. This design paradigm would use the message padding if the input block does not match with the block size or the

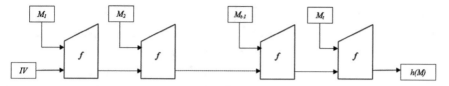

Fig. 2.1 MD construction

multiples of the block size. The first block of the message and the initialization vector is fed into the compression function f. The initialization vector is used to convert the first block of the input message into an unintelligible chunk of a byte array. The output of the compression function f is unpredictable and therefore, performing input-output correlation analysis would be difficult. The output of the compression function is fixed in its length, and it would act as an initialization vector for the successive block to be processed. This process would be extended for all the blocks to be litigated. The output of the ultimate block would stay as the hash value of the given input message. The contemporary digest algorithms like MD5, SHA-160, and SHA2-family imply this technique to produce the message digests.

2.4.2   MD-5 Digest Function

The MD-5 is a 128-bit digest function which implies MD construction principles to produce a message digest. It does MD strengthening on the input message by splitting the message into the 512-bit blocks. Additionally, it reserves the 64 bits of the ultimate block by storing the length of the input string. Therefore, this algorithm could be only used to process a message of length $<2^{64}$ bits. It uses an array of four internal state vectors, namely A, B, C, and D such that each vector remains of size 32 bits. The chain of the internal state vectors would produce a 128-bit value that would serve as an initialization vector for the adjacent block to be processed. The algorithm is operated on 64 rounds and each round is marked by **Ki**. The value of the state vector would change for every round. The MD-5 algorithm operates four auxiliary functions, namely F, G, H, and I. These functions accept three state vectors B, C, and D as input variables. They manipulate the input values at the bit level using AND, OR, NOT, XOR, MOD, and SHIFT operators. The array of state vectors are circularly shifted one position right for each round to be performed. The auxiliary functions are replaced for every 16 rounds and each round would produce a 128-bit intermediate output. The intermediate hash output is later then used as a state vector for the adjacent block to be processed. The intermediate hash value of the ultimate block would be the hash of the given message. The principle of operation for the MD-5 digest function is sketched in Fig. 2.2. The MD-5 digest function was completely broken during the year 2004. Therefore, the application of the MD-5 algorithm is strictly ruled out for cryptographic use.

Fig. 2.2 Operating
principle of MD-5 digest
function

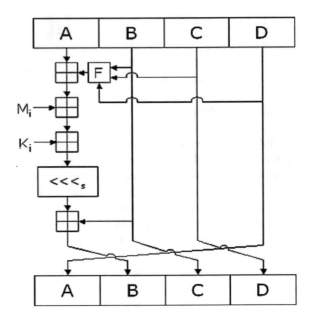

2.4.3 SHA-160 Digest Function

The SHA stands for the secure hash algorithm. It is the first 160-bit digest function produced in the SHA series. It employs the MD construction principle for producing a message digest. This digest function would demand a minimum of 2^{80} iterations to launch a birthday attack and 2^{160} iterations to launch a brute-force attack. Both of these methods are computationally infeasible to perform. Therefore, this digest function was considered as more secure than the MD-5 digest function. The SHA-160 divides the input message into 512-bit blocks. It performs MD strengthening much similar to the MD-5. In contrast to the MD-5, the SHA-160 digest function uses five state vectors, namely A, B, C, D, E, and F such that each state vector is 32-bit in size. The number of rounds has been increased to 80. It uses four auxiliary functions and four keys. The auxiliary function operates solely on B, C, and D vectors and would be changed for every 20 rounds. Similarly, it replaces the key for every 20 rounds. The principle of operation for the SHA-160 digest function is sketched in Fig. 2.3.

The digest function divides every block into 16 words each of 32 bits in size. The words are numbered as W0, W1, W2, ...W15. The digest function calculates Wt for every round $t = 16$ to $t = 79$ using Eq. (2.1).

$$W(t) = S^1 \left(W(t-3) \oplus W(t-8) \oplus W(t-14) \oplus W(t-16) \right) \qquad (2.1)$$

Fig. 2.3 Operating principle of SHA-160 digest function

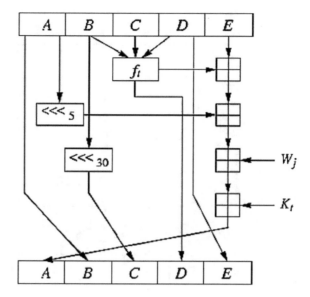

2.4.4   SHA2-256 Digest Function

The SHA-2 family composed of six variants of digest functions, namely SHA2-224/256, SHA2-256, SHA2-224/512, SHA2-256/512, SHA2-384/512, and SHA2-512. Amidst the SHA2 family, the SHA2-256 and SHA2-512 are considered as base versions. The other variants are typically considered as the truncated forms of either SHA2-256 or SHA2-512. The construction principle of SHA2-256 and the SHA2-512 are the same with the sole exception of their block and word sizes. The SHA2-256 operates on 512-bit blocks while the SHA2-512 manipulates the 1024-bit blocks. Similarly, the SHA2-256 allocates the last 64 bits of the ultimate block for the length attribute of the input string, whereas the SHA2-512 shares the last 128 bits for the length of the input string. The word size of SHA2-256 continues 32 bits. In contrast, the word size of SHA2-512 bides 64 bits. The other functional aspects of the SHA2-256 and SHA2-512 are identical. Therefore, this section presents the construction principle of SHA2-256.

The SHA2-256 digest function divides each input block into 16 words, namely W0, W1, W2,...W15 such that each word Wi remains of size 32 bits. It uses 8 state vectors, namely A, B, C, D, E, F, G, and H. The algorithm works on 64 rounds to produce a message digest. It employs 16 words for the first 16 rounds. It calculates the words W16 to W63 using Eq. (2.2). It uses 64 keys, namely K0, K1, K2,...K63 and each key remains a 32 bit constant. The algorithm deploys a unique key for every round to be processed. It employs six logical functions on the input message to be processed. The logical functions implemented in the SHA2-256 are presented at the end.

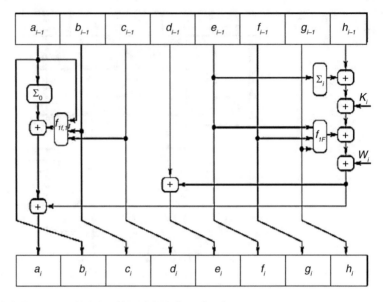

Fig. 2.4 Operating principle of SHA2-256 digest function

1. Ch(x, y, z)=(x ^ y) ⊕ (x ^ z)
2. Maj(x, y, z)=(x ^ y) ⊕ (x ^ z) ⊕ (y ^ z)
3. $\sum_0(x) = S^2(x) \oplus S^{13}(x) \oplus S^{22}(x)$
4. $\sum_1(x) = S^6(x) \oplus S^{11}(x) \oplus S^{25}(x)$
5. $\sigma_0(x) = S^7(x) \oplus S^{18}(x) \oplus R^3(x)$
6. $\sigma_1(x) = S^{17}(x) \oplus S^{19}(x) \oplus R^{10}(x)$

The symbols S^i and R^i present in the logical functions symbolize the circular left shift and right shift operations, respectively. The code snippet of the round function used in the SHA-512 is given at the end. The principle of operation for SHA-256 is sketched in Fig. 2.4.

$$Wi = \begin{cases} mi & \text{for } i < 16 \\ \sigma_1 m_{i-2} + m_{i-7} + \sigma_0 m_{i-15} + m_{i-16} & \text{for } i > 16 \text{ and } i < 64 \end{cases} \quad (2.2)$$

1. PROCEDURE SHA-256
2. BEGIN
3. //Initialize the registers a to h
4. a=H1₀; b=H2₀; c=H3₀; d=H4₀, e=H5₀, f=H6₀, g=H7₀, h=H8₀;
5. For i=0 to N-1
6. {
7. For J=0 to 63
8. {
9. T1=h+$\sum_1$ (e)+Ch(e,f,g)+Kj+wj;
10. T2=$\sum_0$(a)+Maj(a,b,c);

11. h=g; g=f; f=e; e=d+T1; d=c; c=b; b=a; a=T1+T2;
12. }
13. $H1_i=a+H1_{(i-1)}$; $H2_i=b+H2_{(i-1)}$; $H3_i=c+H3_{(i-1)}$; $H4_i=d+H4_{(i-1)}$;
14. $H5_i=e+H5_{(i-1)}$; $H6_i=f+H6_{(i-1)}$; $H7_i=a+H7_{(i-1)}$; $H8_i=a+H8_{(i-1)}$;
15. }
16. // Final hash value is the chain of intermediate digests
17. $H_N=\{H1_N,H2_N,H3_N,H4_N,H5_N,H6_N,H7_N,H8_N\}$
18. END;

2.4.5 SHA3 Digest Function

The SHA3 digest function represents the outcome of the direct competition conducted by NIST in an attempt to find a new standard for the digest functions. The NIST had announced SHA3 as the current standard for the cryptographic digest function during October 2012. The SHA3 differs from SHA2 and the SHA1 in its construction. It used sponge construction to produce a message digest. The SHA3 uses permutation function f to introduce erratic behavior. The width of the function f is marked as b. The digest function divides the message into r bit blocks where r indicates the bit rate. The value of r is selected such that $r < b$. The padding would be applied if the message does not appear as the multiples of r. The capacity is indicated by c calculated by subtracting r from b. The sponge construction works on two levels, namely the absorption phase and squeezing phase. In the absorption phase, a state of b bits is adopted and is initialized with zeros. The r bits from b are XORed with the r bits of the padded message. The resulting b bits are fed into the permutation function f. The result of the f would contain b bits, and it would act as a state bit for the adjacent block to be processed. This process would be prolonged until all the blocks are sued. When the absorption phase is over, the algorithm switches to the squeezing phase. At this phase, the permutation function f is applied on stages until the required number of output bits are produced. If the output bits produced are more than the required size, then first n bits are truncated to produce the final digest. The sponge construct exhibits very strong resistance to the pre-image collision attack. Figure 2.5 illustrates the operating principle of the SHA3 digest function.

2.4.6 Contemporary Digest Functions: A Security Analysis

The attacks on contemporary digest functions prove the digest of size 160 bits does not enough to defend the collision attacks. The partial attacks on SHA2-256 and the SHA3 digest functions prove the security is not relying on the construction principle. The entries of Table 2.1 witness this fact that both RO and the sponge design are prone to partial attacks. The entries also prove that these algorithms would be

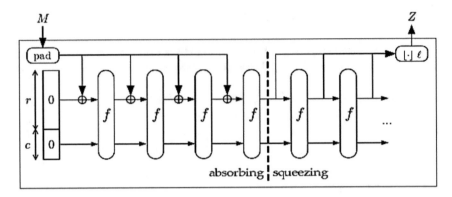

Fig. 2.5 Operating principle of SHA3 digest function

broken soon. The contemporary digest functions excessively imply binary operators like AND, OR, XOR, TRUNC, MOD, NOT, and SHIFT to foresee erratic behavior in the hash output. These operators do not demand much computational power. Therefore, they perform well in terms of efficiency. At the same instant, they are traceable at the block level and for this reason, they are destitute in the perspective of security. The static reservation of the last 64/128 bits of the MD5, SHA-160, and SHA2 family would make the digest functions not to handle the input message of size $>2^{64}/2^{128}$ bits. The dynamic reservation policy would enable them to process a message of any desired length.

2.5   Polynomial Digest

The polynomial function typifies a unidirectional function by its nature and decrypting the polynomial result from the inverse direction remains an arduous task to perform. In the same way, providing a solution for the higher-order polynomial function is a research problem and there is no standard solution available to solve higher-degree polynomial equations (Pan 1997). Therefore, the application of the polynomial in the design of a round function would yield better results from the perspective of security. This part outlines the principles involved in the design of a polynomial digest, the design challenges, and the experimental analysis of the 512-bits polynomial digest function. The digest value is also subjected to confusion and diffusion analysis to examine its stochastic behavior. The digest value is also analyzed for avalanche property, near-collision resistance, and the strict avalanche criteria at the binary level to confirm its stochastic behavior.

2.5.1 Design Challenges

The proposed work deploys a higher-degree polynomial function of degree 64. The digest function would raise a unique 512-bit digest for every distinct input to be processed. The proposed work is hereinafter mentioned as Polynomial Digest Function (PDF). The PDF employs Eq. 3 for the design of block-level round function.

$$P(x) = P_n x^n + P_{n-1} x^{n-1} + P_{n-2} x^{n-2} + \ldots + P_0 x^0 \qquad (2.3)$$

The coefficients of the polynomial are replaced with the block elements of the input string. The polynomial of the form given in Eq. (2.3) represents a unidirectional function and it naturally complies with one-to-one mapping. Therefore, it could be considered as an ideal choice for the design of the digest function. However, Eq. (2.3) could be unapplied directly to the digest function for the following reasons.

- The digest function ordinarily processes arbitrary length input strings without length restrictions. Therefore, it would demand a polynomial function that dynamically adjusts its degree based on the length of the input string. The application of dynamic polynomials would naturally demand tremendous computational power for a very large input string. In contrast, the smaller input strings would force the digest function to produce the output fall short of the required output length.
- The polynomial equation of the form presented in Eq. (2.3) would be more sensitive for the first few bytes of the input elements. This is owing to the presence of intense polynomial powers in Eq. (2.3). They are least sensitive to the ultimate elements of the input string. This would happen when the output of the polynomial is truncated to have a fixed size output. In such circumstances, it would not respond to the changes that happen at the ultimate elements of the input string.

2.5.2 Design Principles

The PDF uses the MD iterative model for the design of a round function. It operates on three levels to generate a message digest as given at the end.

1. MD Strengthening
2. Intermediate hash generation
3. Final hash generation

2.5.2.1 MD Strengthening

MD Strengthening represents level 1 of the hash-generative process. At this point, the preprocessing is applied to the input string to make it suitable for the block-level processing. The PDF processes the input string as 1024-bit blocks and it performs message padding to cause the input string to appear as multiples of the block size. The padding rule of the PDF differs from the contemporary hash algorithms like MD5, SHA-160, and the family of SHA2 digest functions. It employs a dynamic reservation policy for storing the length attribute. Therefore, it does not reserve the last 64/128 bits to store the length attribute of the input string. By choice, it calculates the length value at runtime and it reserves the storage space accordingly. The length attribute of the input string is typically appended after the padding bits of the ultimate block to be processed. A terminal element 0X80 is successively appended to the input string to differentiate the input message from the other control information. It would fill the rest of the block elements with zeros. The modified padding rule would enable the PDF to process the input string without any length restrictions. The message padding P is performed using Eq. (2.4). Figure 2.6 illustrates the MD strengthening of the PDF.

$$P = B - (M + \Phi + 1) MOD B \tag{2.4}$$

At this place, the value M indicates the byte count of the input data and Φ represents its length attribute. The B points the block size, i.e., 128 bytes or 1024 bits.

After the MD strengthening, the PDF applies the initialization vector (IV) to the preprocessed input elements. The IV involves an array of eight elements such that each element remains 128 bits in size. The chain of the vector elements would produce a 1024-bit block. The PDF would modify the IV if the preprocessed input string contains two or more blocks. This would help the PDF to supply distinct IVs for every input block to be processed. To perform this, the distinct elements of the IV are circularly rotated five bit-positions left and the vector array elements are circularly rotated one position right. Figure 2.7 presents the IV elements of the PDF. The application of IV elements would transform the input bytes into an unintelligible chunk of a byte array. This would prevent the cryptanalyst to perform a correlation analysis between input elements and the digest. The response of the PDF for a sample input data after the MD strengthening is presented in Fig. 2.8.

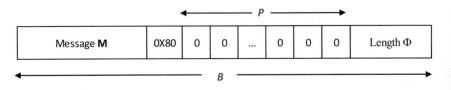

Fig. 2.6 MD strengthening of the PDF

Fig. 2.7 IVs used in the PDF

12127097d0e50429139687921d0c4305
222ac973ef90ef324963300cb671af25
1921bfdb4ad551512a9fa7fcc88aeed1
3171e5b896581ef1fa76c99650faa453
21e77dba903f689736a5807a2dec5092
3ca1f39e420c93728229acbef62a2d92
43658fb59a940f9462cfdbcb2d743746
2c9a8137dceae91d488e0e20c5c9ad32

F:\Papers\Chapter Proposal\Code-512>java PreProcessArrayElements
The Given Input String is :-
(yeast), and a universal solvent (water!) to create a pleasant tasting, psychoactive substance called alcohol. Alcohol is a drug,
similar to many other drugs that every culture known to man has used to expand the consciousness, commune with the gods, or just to catch a buzz. Alcohol
in general, and beer in particular, may have been the driving force behind the eventual civilizing of the beast, man. After all, without fermentable, there is no
alcohol, and without cultivation, and therefore
The length of the Input String is :- 512
The ASCII values of the Input String are :-
13 13 10 40 121 101 97 115 116 41 44 32 97 110 100 32 97 32 117 110 105 118 101 114 115 97 108 32 115 111 108 118 101 110 116 32 40 119 97 116 101
114 33 41 32 116 111 32 99 114 101 97 116 101 32 97 32 112 108 101 97 115 97 110 116 32 116 97 115 116 105 110 103 44 32 112 115 121 99 104 111 97
111 97 99 116 105 118 101 32 115 117 98 115 116 97 110 99 101 32 99 97 108 108 101 100 32 97 108 99 111 104 111 108 46 32 65 108 99 111 104 111 108
108 32 105 115 32 97 32 100 114 117 103 44 13 13 10 115 105 109 105 108 97 114 32 116 111 32 109 97 110 121 32 111 116 104 101 114 32 100 114 117 103
115 32 116 104 97 116 32 101 118 101 114 121 32 99 117 108 116 117 114 101 32 107 110 111 119 110 32 116 111 32 109 97 110 32 104 97 115 32
117 115 101 100 32 116 111 32 101 120 112 97 110 100 32 116 104 101 32 99 111 110 115 99 105 111 117 115 110 101 115 115 44 32 99 111 109 109 117 110 101
32 119 105 116 104 32 116 104 101 32 103 111 100 115 44 32 111 114 32 106 117 115 116 32 116 111 32 99 97 116 99 104 32 97 32 98 117 122 122 46 32 65 108
99 111 104 111 108 32 105 110 32 103 101 110 101 114 97 108 44 32 97 110 100 32 98 101 101 114 32 105 110 32 112 97 114 116 105 99 117 108 97 114 44 32 109
97 121 32 104 97 118 101 32 98 101 101 110 32 116 104 101 32 100 114 105 118 105 110 103 32 102 111 114 99 101 32 98 101 104 105 110 100 32 116 104 101
32 101 118 101 110 116 117 97 108 32 99 105 118 105 108 105 122 105 110 103 32 111 102 32 116 104 101 32 98 101 97 115 116 44 32 109 97 110 46 32 65 102 116
101 114 32 97 108 108 44 32 119 105 116 104 111 117 116 32 102 101 114 109 101 110 116 97 98 108 101 44 32 116 104 101 114 101 32 105 115 32 110 111 32 97 108
99 111 104 111 108 44 32 97 110 100 32 119 105 116 104 111 117 116 32 99 117 108 116 105 118 97 116 105 111 110 44 32 97 110 100 32 116 104 101 114 101 102 111
114 101 32 32 32 32 32
The Length of the data String is : 512
The value of the Input String after preprocessing :-
110 51 -54 -126 74 116 -120 -55 75 30 94 -6 -38 0 -60 -18 54 42 -62 -52 -88 -117 -63 64 -50 89 -124 -115 -31 93 13 -88 64 -53 -46 -35 -121 -90 99 124 50 1
-46 -114 -52 -121 -62 22 57 54 -88 -58 -54 -8 -95 39 76 92 77 -99 -67 15 14 34 58 83 -91 98 -34 -49 7 -1 30 64 116 12 105 -59 22 -23 112 104 86 -9 115 95
-41 11 -61 51 -57 77 -98 125 125 18 -54 -39 -53 75 -99 103 -91 -104 30 -57 24 -12 11 83 13 -68 -11 99 5 -75 63 -45 -75 61 -39 112 119 -35 14 -120 50 44
97 -127 0 -38 -80 -42 -57 91 41 -112 22 -63 -106 17 -88 -63 71 -84 122 102 15 92 89 62 -58 -127 47 63 8 -90 57 -88 -109 35 -109 43 31 -64 -18 54 -45 61
112 -60 35 62 66 -22 -41 -63 -77 -63 -113 82 36 42 -123 16 27 -118 -16 85 -46 -85 104 -43 -40 73 -66 -91 -15 76 126 -21 -128 -28 -127 -65 -70
45 112 45 -66 -16 0 -127 -30 16 97 83 -116 -47 102 14 116 81 -85 106 2 -68 -38 -89 5 -12 51 -95 -94 65 15 -6 32 -93 26 -32 -82 87 -5 28 71 102 -60 -125
44 61 -5 -114 -54 -125 -40 116 85 -15 -124 -7 -111 33 -68 -75 -28 114 -15 93 -38 54 -110 -33 56 71 -55 118 13 -128 -27 -107 -105 22 41 -105 97 -57 -92 40
-54 -56 -107 -77 -21 13 -120 -44 -26 75 -19 70 -14 -85 102 -104 -12 -22 -106 -122 -57 -60 104 -96 -23 89 44 -9 -23 -126 -42 39 125 77 60 -67 -30 -65 -34 -
81 -51 -8 -2 114 64 -5 -38 20 96 124 -34 -112 -13 -117 20 -4 -80 59 13 8 -56 -44 -61 -41 -75 13 -70 -41 62 -30 67 -11 -12 36 81 105 -120 60 96 -38 -111 -
109 -72 43 -90 -122 -98 76 80 -83 32 60 -88 97 99 -74 82 -70 115 -77 114 25 39 107 -33 -87 -12 63 80 -33 97 1 -101 78 -91 -80 106 -34 -7 95 68 114 28 -13
-66 -38 45 -26 36 82 -18 -31 -110 -97 -35 -128 79 -71 111 15 32 109 -88 -110 -72 109 -10 -36 3 41 -42 50 11 84 -20 119 -15 20 -106 -73 107 59 -68 29 37 -
23 119 95 -49 74 -23 -80 17 -65 -121 -19 40 11 -40 -105 -73 -42 14 -65 -17 122 -67 19 -89 46 67 -54 -64 67 122 100 -107 24 95 67 -49 66 116 120 20 52 57
94 101 -49 -1 108 -30 65 -68 -35 -70 66 15 6 -118 -86 -25 72 -73 114 -31 50 4 -101 86 89 4 -27 -22 -103 97 2 -69 -25 -38 97 59 67 -113 -64 -32 -105 49 48
3 -26 -65 -32 108 -97 70 51 -119 86 115 6 45 74 75 -86 -71 87 -21 -3 101 43 17 -15 -65 -42 127 -66 -28 -39 -96 -14 10 -93 49 30 -101 -93 -13 119 45 -
85 -74 -22 12 -26 51 -20 122 44 31 -38 67 43 -45 -114 -118 -39 35 38 29 -27 112 -85 111 -38 -3 16 32 -123 119 63 58 126 -49 58 -45 98 90 90 -38 123 -23
-40 20 102 -62 -62 31 -115 -57 -58 -12 -59 -90 76
The Length of the array after padding 640

Fig. 2.8 Sample response of the PDF after MD strengthening for 512-bit data

2.5.2.2   Intermediate Hash Generation

It represents level 2 of the hash productive process. At this point, the input string is transformed into an unintelligible byte array. Similarly, the size of the input string would be made to match the block size or multiples of the block size. The preprocessed array elements are processed at the block level by consecutively litigating one block at a time. The PDF fixes the block size as 1024 bits. It divides the individual block equally into two halves such that each half would contain 512 bits. The two halves are then processed using the block iterated polynomial function of degree 64. The polynomial function uses distinct values for x to process the array elements of the two halves. The individual halves would produce two intermediate results say R1 and R2. The product of R1 and R2 would produce a digest-seed for the block. The digest-seed typically contains 290 to 310 nibbles. The round function would carve out 256 nibbles by taking 128 nibbles on either direction from the mid position of the digest-seed. The XOR of the two 128 nibbles would form the digest value for the block. If the input string contains more than one block, then the output of the current block is got by XORing it with the previous block output. This process would be extended until all the input blocks are processed. The output of the ultimate block would be the digest of the given input string. Figure 2.9 presents the operating principle of the PDF. Figure 2.10 presents the operating principle of the PDF at the block level.

2.5.2.3 Polynomial Product: A Contemporary Way to Achieve Avalanche Effect

The PDF involves two polynomial functions of degree 64 on every independent block to be processed. During the multiplication operation, every individual digit present in the multiplier naturally changes every significant digit of the multiplicand to produce intermediate results. Let A and B are the multiplicand and multiplier. The number of digits present in the multiplicand and the multiplier is m and n, respectively. The product of A and B would produce n intermediate results. Further, the successive $n - 1$ intermediate results are shifted one position left to the previous intermediate results. After the shift operation, the resultant values are added to produce the digest value of the block. Figure 2.11 illustrates the operating principle of the PDF to generate digests for an input block. The analysis of the multiplication operation proves the midriff of the ultimate result would undergo maximum changes. Therefore, the PDF carves out the midsection of the end product to achieve maximum avalanche response.

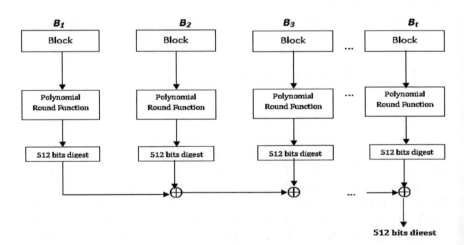

Fig. 2.9 Operating principle of block iterated polynomial digest

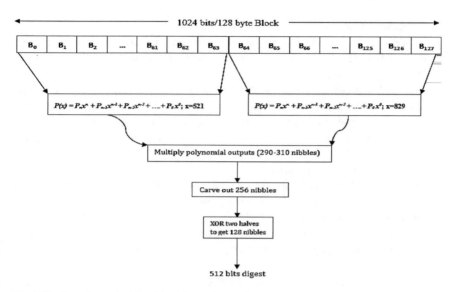

Fig. 2.10 Operating principle of the PDF at the block level

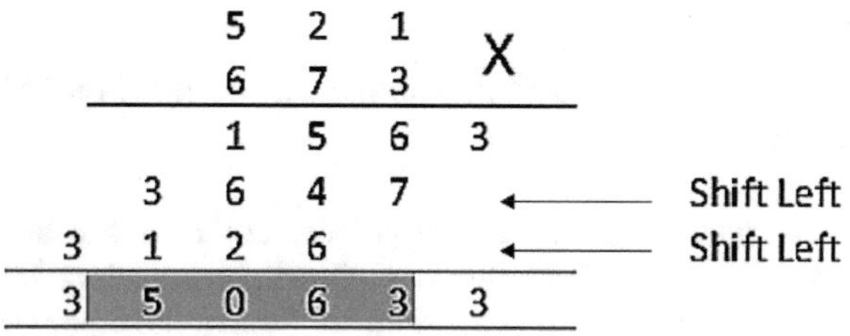

Fig. 2.11 Polynomial products an illustrative diagram for improved avalanche effect

2.5.2.4   Final Hash Generation

This represents the ultimate level of the hash-generative process. At this point, the digest value is inspected for 512-bits/128 nibbles output. Any missing bits would be compensated by prefixing zeros to the digest value.

2.5.3   *Experimental Analysis*

The PDF has been subjected to experimental analysis to investigate its response to the significant security attributes of a cryptographic digest function. The key properties were conveyed at the end for a brief reference. The experiments were conducted using Intel(R) Core(TM) i3 6006U CPU @ 2.00 GHz processor. The digest function was completely devised with JDK 10.0.1 and tested using the Windows-10 64-bit operating system.

1. **Collision resistance**
   This property advocate that, for any two different messages the $x1$ and $x2$, their digest values $H(x1)$ and $H(x2)$ would differ from one another, i.e., $H(x1) \neq \mathrm{H}(x2)$.

2. **Pre-image resistance**
   This property could be referred to as a one-way property. It says, for any input data x, it is straightforward to generate $H(x)$. At the same time, computing x from $H(x)$ would be unusually hard to perform. To state it differently, it is impracticable for a cryptanalyst to recognize a different message x' such that $H(x) = H(x')$.

3. **Second pre-image resistance**
   The property says that for any two distinct messages $x1$ and $x2$, their hash values $H(x1)$ and $H(x2)$ would differ from one another, i.e., $H(x1) \neq H(x2)$.

4. **Avalanche response**
   This property postulates that for every insignificant change in the input string at the binary level would flip the significant number of output bits.

5. **Near-collision resistance**
   This property says that for any two distinct messages the $x1$ and $x2$, their corresponding digest values $H(x1)$ and $H(x2)$ should differ by the least number of bits.

2.5.3.1 Analysis of Collision and Pre-image Resistance (Modifying the Individual Bytes)

Initially, the PDF was put down to inspect the collision resistance, pre-image resistance properties. To conduct this experiment, the distinct input strings of varying length were employed. Every significant byte of the input string was taken and it was modified for all the possible byte values ranging from 0 to 255. The corresponding changes in the hash output were registered and were linked with the reference hash value for possible collision and pre-image collision. More than 4.8 million hash searches were performed to inspect the collision and pre-image collision properties of the PDF. The result of the experiment is presented in Table 2.2. The outcome of the experiment proves the PDF did not produce any collision and pre-image collision and it exhibits a formidable resistance to collision and pre-image collision.

2.5.3.2 Analysis of Collision and Pre-image Resistance (Interchanging the Individual Bytes)

The PDF is once again inspected for collision and pre-image resistance properties through interchanging every individual byte of the input elements with other elements. The changes in the hash output corresponding to every byte interchange were recorded and they were linked with the reference value for a possible hash collision. More than 26.7 million hash searches were performed to inspect the

Table 2.2 Collision and pre-image collision response of the PDF (modifying the byte values)

S. no	No.of input blocks taken	Data size (bytes)	No. of Hash comparisions made (worst case)	No of hash collisions found	Findings
1	<1 Block	32	6630	0	Proposed algorith exhibits
2	1	64	8160	0	strong collision and
3	2	128	16320	0	pre-image resistance
4	4	256	32640	0	properties
5	8	512	65280	0	
6	16	1024	130560	0	
7	24	1536	261120	0	
8	32	2048	391680	0	
9	40	2560	522240	0	
10	48	3072	652800	0	
11	56	3584	783360	0	
12	64	4096	913920	0	
Total number of Hash comparisons made			**4829190**		

collision and pre-image resistance properties of the PDF. The outcome of the experiment is confronted in Table 2.3 for analysis. The result proves the PDF shows formidable resistance to collision and pre-image collision.

2.5.3.3 Analysis of Second Pre-image Resistance

To perform this test, a domain of input strings comprising 100 thousand of elements were formed. The individual element present in the domain differs from one another in size and the content. The PDF is then provided with any two arbitrarily selected distinct samples from the collection of input strings. Their hash outputs were recorded and compared for a possible hash match. In the same way, more than 50,000 comparisons were made with 100 thousand of sample inputs. The outcome of the experiment is confronted in Table 2.4 for analysis. The result proves the PDF did not produce any second pre-image collision. Therefore, the PDF exhibits formidable resistance to the second pre-image collision.

2.5.3.4 Confusion and Diffusion Analysis

Random behavior stays an essential requisite for the digest functions. This behavior not only prevents the cryptanalyst to perform differential analysis and also it significantly increases the security of the digest function. This is because the digest function would behave like a compression function when the input string is larger than the digest value. In such circumstances, the digest function would produce

Table 2.3 Collision and pre-image collision response of the PDF (interchanging the byte values)

S. no	No. of input blocks taken	Data size (bytes)	No.of Hash comparisions made (worst case)	No of Hash collisions found	Findings
1	<1 Block	32	496	0	Proposed algorith has
2	1	64	2016	0	strong collision and
3	2	128	8128	0	pre-image resistance
4	4	256	32640	0	properties
5	8	512	130816	0	
6	16	1024	523776	0	
7	24	1536	1178880	0	
8	32	2048	2096128	0	
9	40	2560	3275520	0	
10	48	3072	4717056	0	
11	56	3584	6420736	0	
12	64	4096	8386560	0	
Total number of Hash comparisons made			26772752		

Table 2.4 Second pre-image collision response of the PDF

S.no	No. of samples taken	No.of Hash comparisons made	Collision count	Findings
1	1000	500	0	The polynomial digest function did not
2	2000	1000	0	produce any second preimage collision
3	3000	1500	0	
4	4000	2000	0	
5	5000	2500	0	
6	6000	3000	0	
7	7000	3500	0	
8	8000	4000	0	
9	9000	4500	0	
10	10000	5000	0	
11	20000	10000	0	
12	30000	15000	0	
Total	**105000**	**52500**	**0**	

hash collisions under the pigeon-hole principle. Therefore, any hash algorithm which performs a rather densification operation would certainly produce hash collision when it is trialed for a brute-force attack. But the intelligent selection of the digest size would help the digest function to overcome this threat. The digest of size 512 bits would demand the cryptanalyst a minimum of 2^{512} iterations to launch a successful brute-force attack. In the same way, it would demand a minimum of 2^{256}

iterations to launch a successful birthday attack. But both these methods are computationally infeasible to perform for the following reasons.

- The time involved in launching these attacks typically demands an extremely extended time.
- The computing power needed to imply these methods on a 512 bits digest function would simply outstrip the capacity of any computing machines available on the planet.

Consequently, the only viable solution convenient for an antagonist to break the PDF is to launch a differential attack. At the same time, launching a differential attack on a system exhibiting random behavior remains an arduous task to perform. In light of the aforesaid fact, the PDF is inspected for its erratic behavior using its avalanche response. To conduct this experiment, a sample input data was taken from the collection of input strings and it was varied a bit at the arbitrarily selected location(s). The resulting changes that happen in the PDF output at the binary level were recorded for analysis. The changes were then ultimately linked with the reference for a possible match at the binary level. Table 2.5 presents the confusion and diffusion analysis at the binary level. The result proves the PDF modifies the output bits evenly across the sheer length of the digest for a single input bit-flip.

The statistical analysis of confusion and diffusion is graphically analyzed to monitor the behavior of the PDF on avalanche property. To conduct this experiment, a sample input string of 256 bytes was chosen, and it was employed to the PDF to produce a 512 bits output. This output was considered as a reference output. At this point, the input string was modified by flipping a bit/bits at the arbitrarily selected location(s). The modified bytes of the input string were then applied to the PDF to produce a 512-bit digest for the modified input. The resulting digest value was equated with the denoted output at the binary level by comparing the bit values using the index positions. The index positions of the changed bits, i.e., unmatched locations, were identified and were stored in a separate array. A graph was plotted using the unmatched index positions to study the avalanche behavior of the PDF. Figure 2.12 presents the unmatched location history of the PDF at the bit level. In the same way, the index positions of the unmatched locations were identified by comparing the nibbles of the modified output with the reference output. Figure 2.13 presents the location history of the matched nibbles of the PDF responses. The result proves the PDF modifies 122 nibbles among the 128 available nibbles.

The results gained from Figs. 2.12 and 2.13 prove the following facts.

- The changes in the output bits are arbitrary.
- The PDF uniformly affects the output bits even for a single bit-flip at the input.
- The output bits are uniformly varied throughout the sheer length of the digest.

Further, Fig. 2.13 demonstrates the modified digest nibbles matched with the reference nibbles only in six locations. It modifies the remaining 122 nibbles for a single bit-flip at the input string. This indicates the avalanche response at the binary level is stochastic, and it produces some unpredictable patterns in the output nibbles. If the input string is modified a bit by varying the location of the input string, then

Table 2.5 Statistical analysis of confusion and diffusion for an input of size 256 bytes

Sample input string	Size = 256 bytes			Avalanche %	Near collision %
S.no	Location	Type	Hash value		
1	NA	e-Version: 1.0 Content-Transfer-Encoding: 7BIT Hi, I wish to have a file uploaded to the hombrew archive. I have previously written to you without a reply. I am including the upload at the end of this message so that you may upload Reference			
		Reference Hash (Binary)	100101101000001101011111101101001001010000001010110110111001011110100011100011000011000110011000001001001110101010101001000110000100111111100010010100101001001001100011001101010001100000 101101101111000000110110111101000011011011110100010111101010010 110101001000010010011100100100010100111110011100110010010101 100100110101111111101100010000100000010111111101011110110100110011001001011110100100010001000001111111110110101010100010111110000010010100110101010100010111110001101011110100101010000010	NA	NA
		Reference hash (Hex)	4b41afb282b79e47aec467049d5488e21fc52c935305b783745ddf844e5ead4834e4854f33e54ac9bfd88 417eb2e426d241d9883fdaa6f0d942a6097c35e8a82		
2	3	Modified Hash (Binary)	011001010011001110110110000110101010110110011001100010001001101111000001011111000 00010101110000100011111011110110111000000110000010000011110110111010000001100 01010101011001100 11110000110010010011010111111111101010101010010010011000000000100 011000010101001011101011100100000 1010110000111011011011001110100011011001010111110 101100011111011101101101001001001110010100 10011111100000011101110010100 100111100110011000011111100100101010100010011101110101001101011000011111101010101010100011110110000101011000	50.39	49.61
		Modified Hash (Hex)	669dc35672370889b18bc1589eec0611392bd8d83156678637fe92891e000460d5d60ac385ed7306a5eb 1eede6749f0819de4cf9879290 8a9 c5e75207ec3b858		

3	Modified Hash (Binary)	110111101100110011110010110101100001000100000001110101010100100100100100100111101011010100000001101101111101010101011101011011010100010010000011110111111100111101011011010101001011101111111101010101010010010011101101011110000010011011010101010100001011111111001100111110000100001010000111100011011000010011010010000110011001100010101010101000011010100100000000000011100000100111110000	49.02	50.98
	Modified Hash (Hex)	de99e5b09076a8448e901b7daf6d07e777736f764eb5d4b9162bd67bfaad3a7506760af7531b5fe7c406fb0 a3ba8a60062c54785603c23643314e0066004384e		
4	Modified Hash (Binary)	100101111011010100100011111110111011011110011001100101111001101101101100010000110011101111110 11100010111101011101010110111011110101010100100011011111111100010100010100001001011111101101 10001110110011100010001011110101011101101 10111101000010000010011001 101011010010110011001110100010110011001100001111100010101010101010001100100100110000010011110110 01101011010100001100011110011111111001100001011101101011010100110100010101010100110110001000 01101101010010000001000000100100101010000010010001000010001011111000001101010001000	50.98	49.02
	Modified Hash (Hex)	97698fdddf332f36d11df7175daef5a310cd455425d8ec43bf29e41b21dfc67b8b12eb6e8f3382e2cc69f6f1 83c10cd5364e60f2a1233688202a1084f0882728		

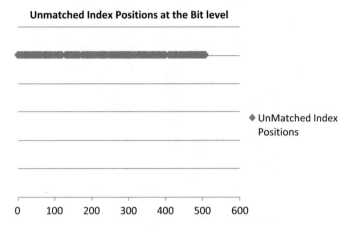

Fig. 2.12 Index positions of the unmatched bits

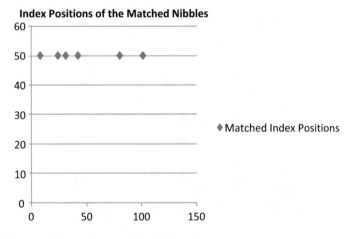

Fig. 2.13 Index positions of the matched nibbles

the PDF would produce a unique output. These distinct responses are random and therefore they are unusually complex to be equated with one another. Therefore, performing differential analysis against the PDF is the most grueling task.

2.5.3.5 Analysis of Avalanche Response

To analyze the avalanche behavior of the PDF, the proposed design is supplied with diversified input strings such that the input strings differ from one another in their content and the length. The input string is modified by flipping the input bits as given in Table 2.6. The bitwise comparisons were performed at the binary level using the index positions of the digest values produced. Every modified digest value

Table 2.6 Avalanche response of the PDF for <4 K data

S.No	No.of input blocks taken	Data size (bytes)	Avalanche response of the PDF for the data modification ranges from 1 bit to 32 bytes applied on the input data of sizes ≤4 K													Average Avalanche response for the data changes ≤32 Bytes
			1 bit	2 bit	3 bit	4 bit	5 bit	6 bit	7 bit	1 B	2 B	4 B	8 B	16 B	32 B	
1	<1 Block	46	49.85	50.01	49.67	50.18	49.59	50.22	50.03	49.73	50.06	49.92	50.09	50.04	49.99	49.95
2	1	64	49.96	50.16	50.43	50.04	50.21	50.22	50.38	50.27	50	50.07	49.92	50.02	49.96	50.13
3	2	128	49.78	49.88	50.06	50.04	50.05	50.08	50.13	50.08	50.13	49.93	49.83	50.11	49.9	50.00
4	4	256	49.99	50.06	49.87	50.2	49.95	49.95	50.01	50.03	49.91	49.83	50.01	50.04	49.96	49.99
5	8	512	50.04	50.09	49.87	50.01	50.11	50.3	50.2	49.65	49.97	49.78	50	50.08	49.88	50.00
6	16	1024	50.02	49.97	49.97	50.11	49.78	49.93	49.98	49.95	50.04	49.93	50.11	50.02	50.05	49.99
7	24	1536	50.15	50	49.92	50.07	50.09	50.18	50.05	50.02	50.01	50.14	49.96	49.99	49.9	50.04
8	32	2048	49.96	50.04	49.93	50.03	50.1	49.78	49.98	50.08	50.01	50.11	49.96	50.03	49.79	49.98
9	40	2560	49.8	50.02	49.96	50.05	49.97	49.97	50.02	49.97	49.87	49.9	50.07	49.93	50.14	49.97
10	48	3072	50.06	50.19	49.83	49.91	49.9	49.88	50.01	49.92	50	49.97	50.16	50.06	49.78	49.97
11	56	3584	49.96	50.01	49.85	50.1	50.17	50.05	50	50.12	50.02	49.98	50.02	50.08	49.88	50.02
12	64	4096	50.13	50.02	50.02	49.94	50.05	49.97	49.81	50.04	50.08	50.02	49.99	49.9	50.04	50.00
Average Avalanche Response for Data < 4K			**49.98**	**50.04**	**49.95**	**50.06**	**50.00**	**50.04**	**50.05**	**49.99**	**50.01**	**49.97**	**50.01**	**50.03**	**49.94**	**50.00**

is equated with the denoted value to study the avalanche behavior of the PDF. Table 2.6 presents the outcome of this test. The typical entries of Table 2.6 represent the average value of 500 samples. The results prove the average avalanche response of the proposed design at the binary level is 50. The result proves the PDF meets the strict avalanche criteria. Therefore, launching a differential attack against the PDF remains an arduous task.

Figure 2.14 presents the graphical response of the PDF resulting from a single bit-flip performed on distinct input strings. Here, the distinct input strings of varying length are bit-flipped at the arbitrarily selected locations. The change in the output nibbles is compared with the reference output at the binary level. The entries of Fig. 2.14 are derived from the mean of 500 hash comparisons. The result proves the PDF modifies the output bits from 49.78% to 50.15%. The average avalanche response of the PDF for a single bit-flip is 49.98%. Figure 2.15 presents the graphical outcome of this test performed on 512 bytes input data. At this point, the average avalanche response of the PDF marginally increased to 50.00%. In the same way, Fig. 2.16 showcases the avalanche response of the PDF for 4 K data. The result proves the PDF maintains a 50% avalanche response for 4 K data as well. Figure 2.17 presents the outcome of this test for the data modification ranges from 1 bit to 32 bytes applied on various input data of size ≤4 K. The results prove the PDF maintains the average avalanche response slightly above 50% for 64 bytes, 128 bytes, 1.5 K, 3.5 K, and 4 K input data. Table 2.6 entries witness the PDF consistently modifies 50% of the end product randomly across the sheer length of the digest irrespective of the length of the input string employed and the number of input bits/bytes modified. Therefore, the proposed design satisfies the strict avalanche criteria of the strong cryptographic digest function. In light of the outcome of the experimental results, launching a differential attack or any block-level attacks on the PDF would be extremely hard.

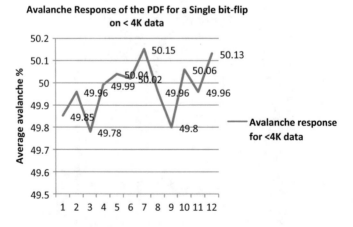

Fig. 2.14 Avalanche response of PDF for a single bit-flip on <4 K data

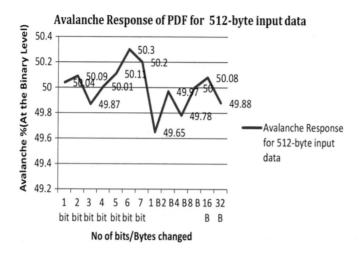

Fig. 2.15 Avalanche response of PDF on 512-byte data

The proposed design was examined for its behavior on the output nibbles in response to the avalanche effect at the binary level. To conduct this experiment, the PDF was supplied with distinct input strings such that the input strings differ from one another in their content and the length. The input string is modified by either flipping the input bits or inserting some random bytes at the arbitrarily selected locations. The responses of the PDF were recorded, and their digest nibbles were equated with the denoted. This is performed to measure the change of output symbols in response to the avalanche effect at the binary level. Table 2.7 presents the outcome of this test. The typical entries of Table 2.7 are calculated by taking an average of 500 samples. The result proves the average change of output nibbles in response to the avalanche effect is 93.75%. The result evidences the PDF changes 120 nibbles among the available 128 nibbles in response to a tiny input change. The result showcases the near-random response of the PDF. Therefore, performing a correlation analysis against PDF digests is extremely hard.

Figure 2.18 presents the average change of output nibbles for a single bit-blip on the input data of sizes ≤ 4 K. The analysis of Table 2.7 entries evidences the average change of output nibbles for a single input bit-flip is 93.69% and it ranges from 93.47% to 93.92%. Figure 2.19 presents the response of the PDF corresponding to the changes of an input data of size 46 bytes. The input string is modified between 1 bit and 32 bytes. The result proves the PDF modifies the output nibbles between 93.41% and 94.12%. The result also evidences the PDF consistently modifies 93.74% of the output nibbles. Figure 2.20 presents the responses of the PDF on the input data of sizes 1 K and 4 K. The input data is modified between 1 bit to 32 bytes and the corresponding changes in the digest values are recorded for the analysis. The results prove the PDF modifies the output nibbles between 93.45% and 93.9%. Figure 2.21 presents the outcome of this test on the input data of sizes between 46 bytes and 4 K. Each of the input samples is modified between 1 bit and 32 bytes.

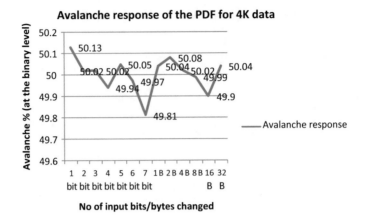

Fig. 2.16 Avalanche response of PDF on 4 K data

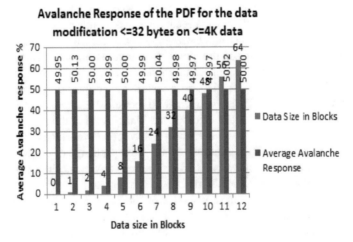

Fig. 2.17 Avalanche response of PDF for the data changes of ≤32 bytes on ≤4 K data

The result showcases the average of change of output nibbles of the PDF lies between 93.65% and 93.85%. The experimental results prove the consistent performance of the PDF in changing the output nibbles irrespective of factors like the number of bits/bytes changed and the size of the input data. As an average, the PDF almost changes every output nibbles for even a tiny input change in the input data. Therefore, launching a differential attack would become unusually hard against the PDF.

Table 2.7 Avalanche response of the PDF for <4 K data

S.no	No.of input blocks taken	Data size (bytes)	Average change of output nibbles due to the alanche effect of the PDF for the data modification ranges from 1bit to 32 bytes applied on the input data of sizes ≤4 K													Average change of output nibbles due to the Avalanche response for the data changes ≤32 Bytes
			1 bit	2 bit	3 bit	4 bit	5 bit	6 bit	7 bit	1 B	2 B	4 B	8 B	16 B	32 B	
1	<1 Block	46	93.62	93.59	94.12	93.86	93.94	94.03	93.41	93.41	93.75	93.66	93.74	93.78	93.7	93.74
2	1	64	93.75	93.54	94.16	94.07	94.11	93.49	94.08	93.44	93.75	93.94	93.68	93.58	93.95	93.81
3	2	128	93.47	93.29	93.93	93.62	93.74	93.63	93.82	93.54	93.7	93.75	93.66	93.61	93.67	93.65
4	4	256	93.69	93.57	93.65	94.03	93.4	95.45	93.98	93.85	93.67	93.7	93.61	93.79	93.7	93.85
5	8	512	93.64	93.63	93.67	93.68	93.75	93.5	93.82	93.68	93.65	94	93.93	93.81	93.87	93.74
6	16	1024	93.6	93.66	93.75	93.72	93.66	93.75	93.45	93.78	93.79	93.86	93.74	93.76	93.88	93.72
7	24	1536	93.92	93.78	93.78	93.67	93.89	93.82	93.77	93.56	93.74	93.79	93.7	93.9	93.72	93.77
8	32	2048	93.79	93.69	93.9	93.87	94	93.83	93.92	93.75	93.57	93.61	93.78	93.67	93.78	93.78
9	40	2560	93.65	93.63	93.69	93.76	93.81	93.92	93.78	93.83	93.8	93.85	93.8	93.71	93.66	93.76
10	48	3072	93.78	94.02	93.86	93.68	93.89	93.84	93.75	93.8	93.8	93.78	93.77	93.75	93.63	93.80
11	56	3584	93.72	93.64	93.65	93.85	93.78	93.72	93.64	93.59	93.64	93.74	93.82	93.7	93.75	93.71
12	64	4096	93.59	93.81	93.73	93.64	93.53	93.69	93.88	93.81	93.63	93.67	93.9	93.82	93.53	93.71
Average Avalanche Response for Data <4K			93.69	93.65	93.82	93.79	93.79	93.89	93.78	93.67	93.71	93.78	93.76	93.74	93.74	93.75

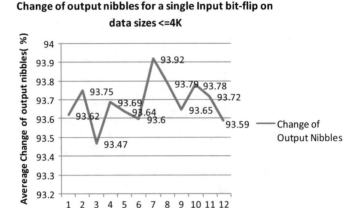

Fig. 2.18 The response of PDF for a single bit-flip on the input data of size ≤4 K

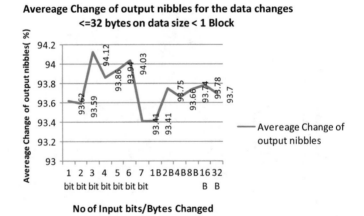

Fig. 2.19 The PDF response for the data changes between 1 bit and 32 bytes on a 46 bytes input data

2.6 Analysis of Near-Collision Response

To analyze the near-collision response of the PDF, the proposed design is inspected with diversified input strings of varying length and the content. The input string is modified between 1 bit and 32 bits at the arbitrarily selected locations. Table 2.8 presents the outcome of the PDF responses for this experiment. The typical entries of the table are calculated by performing an average of 500 samples. The responses of the PDF were stored and equated with the denoted value for a possible match at the binary level. The comparisons were performed by using the index positions of

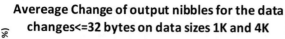
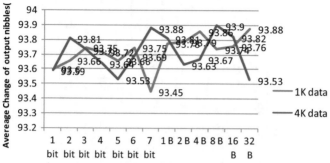

Fig. 2.20 Response of PDF for the data changes between 1 bit and 32 bytes on 1 K and 4 K data

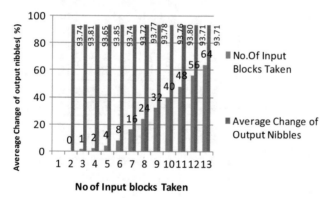

Fig. 2.21 The PDF response for the data changes ≤32 bytes on the input data of sizes ≤4 K

the digest at the binary level. The results prove the average near-collision response of the proposed design at the binary level is 50. This happens because the changes performed on the output nibbles are arbitrary and the probability for changing the binary bit is 0.5. The PDF output is also analyzed for the near-collision effect on the output nibbles. Table 2.9 presents the outcome of this experiment. The result showcases the average matched nibbles are 6.25 and it recapitulates the fact that the changes happened in the digest output are unbiased and uniform throughout the digest length. Thereof, performing a differential analysis against the PDF would demand a back-breaking effort from an attacker.

Table 2.8 Near-collision response of the PDF for the data changes ≤32 bytes on ≤4 K data

S.no	No.of input blocks taken	Data size (bytes)	Near collision response of the PDF for the data modification ranges from 1 bit to 32 bytes applied on the input data of sizes ≤4 K													Average near collision response for the data changes ≤32 Bytes
			1 bit	2 bit	3 bit	4 bit	5 bit	6 bit	7 bit	1 B	2 B	4 B	8 B	16 B	32 B	
1	<1 Block	46	50.15	49.99	50.33	49.82	50.41	49.78	49.97	50.27	49.94	50.08	49.91	49.96	50.01	50.05
2	1	64	50.04	49.84	49.57	49.96	49.79	49.78	49.62	49.73	50	49.93	50.08	49.98	50.04	49.87
3	2	128	50.22	50.12	49.94	49.96	49.95	49.92	49.87	49.92	49.87	50.07	50.17	49.89	50.1	50.00
4	4	256	50.01	49.94	50.13	49.8	50.05	50.05	49.99	49.97	50.09	50.17	49.99	49.96	50.04	50.01
5	8	512	49.96	49.91	50.13	49.99	49.89	49.7	49.8	50.35	50.03	50.22	50	49.92	50.12	50.00
6	16	1024	49.98	50.03	50.03	49.89	50.22	50.07	50.02	50.05	49.96	50.07	49.89	49.98	49.95	50.01
7	24	1536	49.85	50	50.08	49.93	49.91	49.82	49.95	49.98	49.99	49.86	50.04	50.01	50.07	49.96
8	32	2048	50.04	49.96	50.07	49.97	49.9	50.22	50.02	49.92	49.99	49.89	50.04	49.97	50.21	50.02
9	40	2560	50.2	49.98	50.04	49.95	50.03	50.03	49.98	50.03	50.13	50.1	49.93	50.07	49.86	50.03
10	48	3072	49.94	49.81	50.17	50.09	50.1	50.12	49.99	50.08	50	50.03	49.84	49.94	50.22	50.03
11	56	3584	50.04	49.99	50.15	49.9	49.83	49.95	50	49.88	49.98	50.02	49.98	49.92	50.12	49.98
12	64	4096	49.87	49.98	49.98	50.06	49.95	50.03	50.19	49.96	49.92	49.98	50.01	50.1	49.96	50.00
Average Avalanche Response for Data < 4 K			**50.03**	**49.96**	**50.05**	**49.94**	**50.00**	**49.96**	**49.95**	**50.01**	**49.99**	**50.04**	**49.99**	**49.98**	**50.06**	**50.00**

Table 2.9 The effect of near-collision response on the output nibbles for the data changes ≤32 bytes on ≤4 K data

S. no	No.of input blocks taken	Data size (bytes)	Average matching of output nibbles due to the alanche effect of the PDF for the data modification ranges from 1 bit to 32 bytes applied on the input data of sizes ≤4 K													Average matching of output nibbles due to the Avalanche response for the data changes ≤32 Bytes
			1 bit	2 bit	3 bit	4 bit	5 bit	6 bit	7 bit	1 B	2 B	4 B	8 B	16 B	32 B	
1	<1 Block	46	6.38	6.41	5.88	6.14	6.06	5.97	6.59	6.59	6.25	6.34	6.26	6.22	6.3	6.26
2	1	64	6.25	6.46	5.84	5.93	5.89	6.51	5.92	6.56	6.25	6.06	6.32	6.42	6.05	6.19
3	2	128	6.53	6.71	6.07	6.38	6.26	6.37	6.18	6.46	6.3	6.25	6.34	6.39	6.33	6.35
4	4	256	6.31	6.43	6.35	5.97	6.6	4.55	6.02	6.15	6.33	6.3	6.39	6.21	6.3	6.15
5	8	512	6.36	6.37	6.33	6.32	6.25	6.5	6.18	6.32	6.35	6	6.07	6.19	6.13	6.26
6	16	1024	6.4	6.34	6.25	6.28	6.34	6.25	6.55	6.22	6.21	6.14	6.26	6.24	6.12	6.28
7	24	1536	6.08	6.22	6.22	6.33	6.11	6.18	6.23	6.44	6.26	6.21	6.3	6.1	6.28	6.23
8	32	2048	6.21	6.31	6.1	6.13	6	6.17	6.08	6.25	6.43	6.39	6.22	6.33	6.22	6.22
9	40	2560	6.35	6.37	6.31	6.24	6.19	6.08	6.22	6.17	6.2	6.15	6.2	6.29	6.34	6.24
10	48	3072	6.22	5.98	6.14	6.32	6.11	6.16	6.25	6.2	6.2	6.22	6.23	6.25	6.37	6.20
11	56	3584	6.28	6.36	6.35	6.15	6.22	6.28	6.36	6.41	6.36	6.26	6.18	6.3	6.25	6.29
12	64	4096	6.41	6.19	6.27	6.36	6.47	6.31	6.12	6.19	6.37	6.33	6.1	6.18	6.47	6.29
Average Avalanche Response for Data <4 K			**6.32**	**6.35**	**6.18**	**6.21**	**6.21**	**6.11**	**6.23**	**6.33**	**6.29**	**6.22**	**6.24**	**6.26**	**6.26**	**6.25**

2.6.1 Analysis of Input to Output Distribution

The proposed design is examined for input to output distribution. The digest function should uniformly map the input ASCII values into the hexadecimal output digits. The uniform distribution of ASCII values would prevent the cryptanalyst not to perform a correlation analysis between input and output. Figure 2.22 presents the distribution of ASCII values for a given 1 K input data. Figure 2.22 illustrates that the input values are distributed as clusters for the ASCII values between 97 and 122. This was scheduled for the reason that most of the input elements seem to be typically the lower-case alphabets. Therefore, they were distributed closely between the aforesaid range. Figure 2.23 presents the output distribution of PDF of a sample 1 K input data. Unlike the input distribution, the PDF uniformly distributes the hexadecimal digits to the sheer length of the message digest. The distribution of hexadecimal output would completely change for a tiny bit-flip happening at the input. The results prove the mapping of input ASCII values to hexadecimal output digits is random. Therefore, it would strongly prevent the cryptanalyst to perform correlation analysis between input and output values.

2.6.2 Runtime Analysis

The PDF is subjected to comparative analysis on runtime performance with the standard digest functions. The average runtime response of the PDF for data size ≤4 K is 59.52 ms. The results prove the response time of the PDF is directly proportional to the size of the input data. It happens because the intense polynomial powers on the increased number of blocks would demand excess clock cycles from

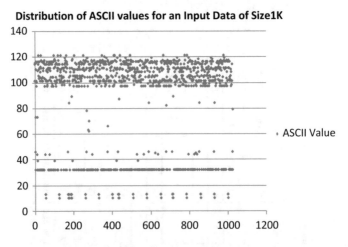

Fig. 2.22 The sample distribution ASCII values for a given 1 K input data

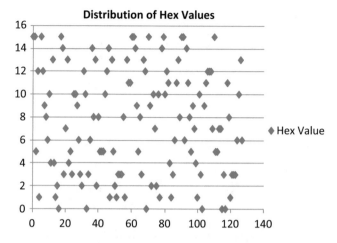

Fig. 2.23 The sample distribution hexadecimal values on the PDF output

the processor. Table 2.10 presents the runtime response of the PDF for the data size ≤4 K. The typical entries of the table are calculated by performing an average of 50 samples.

2.7 Discussion

The security of the digest function is directly proportional to the digest size. RC Merkle insisted an algorithm would demand a minimum of 2^{56} iterations to survive from a birthday attack. But the successful attacks on MD5 and SHA-160 algorithms proved that RC Merkle was wrong. Today the digest of size 160 bits (This would demand 2^{80} iterations to launch a birthday attack.) does be insufficient to defend the cryptographic attacks. The contemporary algorithms like SHA2-224/256, SHA2-256, SHA2-224/512, SHA2-256/512, SHA2-384/512, and SHA2-512 were keen on increasing their digest sizes to enhance their security. The family of SHA3 also follows a similar strategy in fixing their digest size. The aforesaid fact is undisputed to some extent. The SHA2 and SHA3 algorithms were fully unbroken. But the partial attacks on the families of SHA2 and SHA3 raise concerns about this belief. The SHA2 digest function was broken up to 52 rounds and the SHA3 algorithm was broken up to 5 rounds. The precise analysis of the aforesaid partial attacks reveals the fact that the attacks did not solely rely on the digest size but also contingent upon some other factors.

The SHA2 family adopts the RO model for producing a message digest. Contrarily, the SHA3 family uses sponge principles for the design of digest function. It was believed that sponge principles would provide more security than the RO paradigm. This was the reason the SHA3 was declared as the modern standard for the cryptographic digest function. But the partial attacks on the SHA2 and SHA3 digest functions prove the attacks were independent of the type of construction used.

Table 2.10 Runtime comparison of PDF with standard digest functions

S. no	No.of input blocks taken	Data size (bytes)	Runtime response of the digest functions in milliseconds										
			MD5	SHA-160	SHA2-224	SHA2-256	SHA2-384	SHA2-512	SHA3-224	SHA3-256	SHA3-384	SHA3-512	PDF-512
1	< 1 Block	46	1	0.36	0.68	0.28	0.56	0.36	17.64	0.68	0.6	0.28	12.56
2	1	64	1.56	0.36	1.12	0.72	0.64	0.52	21.36	1.32	0.2	0.4	17.2
3	2	128	0.84	0.56	0.88	0.36	0.72	0.4	21.32	0.6	1.32	0.36	24.6
4	4	256	1.28	0.72	0.96	1.2	0.8	0.84	20.6	1.48	0.88	0.6	27.72
5	8	512	1.96	0.24	1.56	0.68	0.8	0.52	16.56	0.72	0.6	0.4	40.68
6	16	1024	1.08	0.44	1.04	1.36	1.28	0.6	19.68	1	0.72	0.72	57.8
7	24	1536	1.76	0.64	1.32	0.72	0.8	1.2	16.88	0.88	0.48	0.76	70.36
8	32	2048	1.28	0.44	0.84	1.04	0.56	0.84	14.92	0.56	0.64	0.8	68
9	40	2560	1.44	0.8	1.68	1.08	1.52	0.92	18.88	1.16	1.08	0.88	99.76
10	48	3072	1.4	0.6	1.16	0.92	0.84	1.28	18.36	0.48	0.72	1	90.44
11	56	3584	1.64	0.48	1.64	0.72	0.76	1.32	17.84	0.88	1.2	1.12	95.76
12	64	4096	1.84	0.6	1.96	1.12	0.88	0.84	16.12	1.24	0.72	1.64	109.32
Average runtime response in milliseconds			**1.42**	**0.52**	**1.24**	**0.85**	**0.85**	**0.80**	**18.35**	**0.92**	**0.76**	**0.75**	**59.2**

The contemporary digest algorithms like MD5, SHA-160, SHA2 family, and SHA3 family use bitwise operators, namely AND, OR, MOD, XOR, TRUNC, and NOT operators to produce a message digest. These operators help the digest function to achieve a quick response time. Simultaneously, these operators are traceable from the inverse direction. The successful attacks on MD5 and SHA-160 algorithms and the partial attacks on SHA2 and SHA3 digest functions recapitulate this fact. Therefore, they perform poorly in the perspective of security.

The proposed design reckons the aforesaid security vulnerabilities into account in the design of the digest function. It fixes the digest size as 512 bits. Therefore, this would demand a minimum of 2^{256} iterations for an antagonist to definitively establish a birthday attack. Similarly, it would demand 2^{512} iterations to launch a brute-force attack. However, both these methods are computationally infeasible to perform.

The proposed design involves a higher-degree polynomial function for the design of round function at the block level. The polynomial function is a natural one-way function. There is no standard solution convenient to solve a polynomial equation with the degree 64. In the same way, decoding a polynomial output from the opposing direction remains a computationally infeasible task. Therefore, the application of polynomial function at the block level would naturally prevent block-level attacks like direct attack, forward attack, permutation attack, backward attack, and fixed-point attack. Besides, the proposed design uses an XOR operator on polynomial products to produce the intermediate hash values. It never uses other bitwise operators at the block level. Therefore, tracing the PDF output from the inverse direction is unusually hard to perform.

Randomness stays the desirable property for the cryptographic digest function. The experimental results prove the average avalanche response of the PDF at the binary level makes up 50%. Therefore, the proposed design meets the strict avalanche criteria suggested by Webster et al. The experimental results on the effect of an avalanche on output nibbles is 93.75%. The PDF modifies 120 output nibbles among the 128 available nibbles. The results showcase that the PDF consistently performs well in modifying the output nibbles even for a tiny bit-flip at the input. The effect of near-collision response on the output nibbles is 6.25%. These results showcase the near-random behavior of the PDF. Therefore performing a differential analysis against the proposed design is extremely hard.

The runtime performance of the PDF is substandard, and it linearly grows with the data size. Table 2.10 entries prove the average runtime response of the PDF amounts to 59.52 ms. This slow performance was happened because of the deployment of intensive polynomial powers at the round function. The results prove the PDF produces a more satisfactory response for smaller data, i.e., ≤256 bytes. Considering the nature of the application and the significance of security, this delay is acceptable. Therefore, the proposed polynomial digest function might be viewed as a constructive alternative for the contemporary digest functions from the perspective of security.

2.8 Conclusion

The proposed work suggests a polynomial based block-chain digest function to address the security vulnerabilities of the standard digest functions. The chaotic hash function could be considered as an alternative for the polynomial based block-chain design as it inherently uses the polynomial function at the cell level. But the number of iterations performed at the cell level to achieve the chaotic response naturally increases the response time of the digest function. Contrarily, this work suggests the diligence of polynomial function directly at the block for the design of round function at the block level. Therefore, this design paradigm would help the digest function to produce quicker response time than the chaotic alternatives without jeopardizing the security. The experimental results with more than 31.5 million hash searches on collision resistance, pre-image resistance, and second pre-image resistance prove the PDF is a provably secure digest function. The results on avalanche response, near-collision resistance, and statistical analysis of confusion and diffusion prove the block-chain based polynomial function meets the strict avalanche criteria. The results also prove the PDF consistently produces near-random response irrespective of the length of the input string and the number of bits/bytes changed. This would certainly help the PDF to exhibit strong resistance to the differential analysis. Therefore, the block-chain based polynomial digest could be considered as a constructive alternative for the contemporary keyless digest functions from the perspective of security.

References

A.F. Webster, S.E. Tavares, On the design of S-boxes, in *Conference on the Theory and Application of Cryptographic Techniques*, (Springer, Berlin, Heidelberg, 1985)

S. Al-Kuwari, J.H. Davenport, R.J. Bradford, Cryptographic hash functions: Recent design trends and security notions, in *IACR Cryptology ePrint Archive 2011*, (2011), p. 565

G. Bertoni et al., Keccak specifications, in *Submission to Nist (Round 2)*, (2009), pp. 320–337

G. Bertoni et al., Keccak, in *Annual International Conference on the Theory and Applications of Cryptographic Techniques*, (Springer, Berlin, Heidelberg, 2013)

S.-j. Chang et al., Third-round report of the SHA-3 cryptographic hash algorithm competition, in *NIST Interagency Report 7896*, (2012), p. 121

J.-S. Coron et al., Merkle-Damgård revisited: How to construct a hash function, in *Annual International Cryptology Conference*, (Springer, Berlin, Heidelberg, 2005)

I.B. Damgård, Collision free hash functions and public key signature schemes, in *Workshop on the Theory and Application of of Cryptographic Techniques*, (Springer, Berlin, Heidelberg, 1987)

N. Diarra, D. Sow, A.Y.O.C. Khlil, On indifferentiable deterministic hashing into elliptic curves. Eur. J. Pure Appl. Math. **10**(2), 363–391 (2017)

J.B. Kam, G.I. Davida, Structured design of substitution-permutation encryption networks. IEEE Trans. Comput. **10**, 747–753 (1979)

D. Khovratovich, C. Rechberger, A. Savelieva, Bicliques for preimages: Attacks on Skein-512 and the SHA-2 Family, in *International Workshop on Fast Software Encryption*, (Springer, Berlin, Heidelberg, 2012)

Y. Li, G. Ge, D. Xia, Chaotic hash function based on the dynamic S-box with variable parameters. Nonlinear Dynam. **84**(4), 2387–2402 (2016)

H. Liu, A. Kadir, J. Liu, Keyed hash function using hyper chaotic system with time-varying parameters perturbation. IEEE Access **7**, 37211–37219 (2019)

U. Maurer, R. Renner, C. Holenstein, Indifferentiability, impossibility results on reductions, and applications to the random oracle methodology, in *Theory of cryptography conference*, (Springer, Berlin, Heidelberg, 2004)

R.C. Merkle, One way hash functions and DES, in *Conference on the Theory and Application of Cryptology*, (Springer, New York, NY, 1989)

V.Y. Pan, Solving a polynomial equation: Some history and recent progress. SIAM Rev. **39**(2), 187–220 (1997)

B. Preneel, The state of cryptographic hash functions, in *School organized by the European Educational Forum*, (Springer, Berlin, Heidelberg, 1998)

T. Ristenpart, H. Shacham, T. Shrimpton, Careful with composition: Limitations of the indifferentiability framework, in *Annual International Conference on the Theory and Applications of Cryptographic Techniques*, (Springer, Berlin, Heidelberg, 2011)

S.K. Sanadhya, P. Sarkar, New collision attacks against up to 24-step SHA-2, in *International Conference on Cryptology in India*, (Springer, Berlin, Heidelberg, 2008)

J.S. Teh, K. Tan, M. Alawida, A chaos-based keyed hash function based on fixed point representation. Clust. Comput. **22**(2), 649–660 (2019)

X. Wang, Y. Hongbo, How to Break MD5 and other hash functions, in *Annual International Conference on the Theory and Applications of Cryptographic Techniques*, (Springer, Berlin, Heidelberg, 2005)

X. Wang et al., Collisions for hash functions MD4, MD5, HAVAL-128 and RIPEMD, in *IACR Cryptology ePrint Archive 2004*, (2004), p. 199

X. Wang, Y.L. Yin, H. Yu, Finding collisions in the full SHA-1, in *Annual International Cryptology Conference*, (Springer, Berlin, Heidelberg, 2005)

Chapter 3
Collaborative Approaches for Security of Cloud and Knowledge Management Systems: Benefits and Risks

N. Jayashri and K. Kalaiselvi

3.1 Introduction

In later stage, various papers identified with the knowledge management and knowledge management system are regardless. Cognizance is a concern of anybody or anything, for representative, real ingredients, information, pictures, or aptitudes, which is caught by methods for trip or getting ready through observing, hearing, or prize. Additionally, appreciation can suggest a speculative or reasonable belief of a locus of study. In improver, it will in general be comprehended or conveyed. Unequivocal capacity is information that has been verbalized, arranged, and saved in sure media. Furthermore, it might be at once passed along to others. Regardless, deduced understanding is such a knowledge that is difficult to switch (Dave et al. 2013). Affiliations need knowledge management for finding, mapping, gathering, filtering data, creating up new information, producing non-open perception into shared ability resources, comprehension, and finding, and including the cost of records to produce data. The three central processes of capacity association are making to make sure about, fitness sharing, and getting used. Appropriated figuring when used, does now not merely lay forth an increasingly noticeable effect on the scientific discipline (Sultan 2013). These functions can be startlingly sorted out, provisioned, applied and decommissioned, and scaled up or down; presenting for an on-demand utility-like event of an assignment and use service dispatching plans of appropriated figuring are extending day by methods for day (Arpaci 2017). Knowledge management moreover has created to be as the key establishment of earnestness and companion's execution. The press to control understanding inside a

N. Jayashri (✉) · K. Kalaiselvi
Vels Institute of Science, Technology and Advanced Studies (VISTAS),
Pallavaram, Chennai, India
e-mail: kalairaghu.scs@velsuniv.ac.in

© The Author(s), under exclusive license to Springer Nature Switzerland AG 2021   57
A. Bhardwaj, V. Sapra (eds.), *Security Incidents & Response Against Cyber Attacks*,
EAI/Springer Innovations in Communication and Computing,
https://doi.org/10.1007/978-3-030-69174-5_3

living class of action is decisive. The nonappearance of progression will clear the association's general execution decrease, while steady improvement will make the affiliation show further growth. Starting at now the use of science as a contrivance of the knowledge organization has created to be a level out factor in an organization and has given various ideal conditions to the knowledge (Aksoy and Algawiaz 2014). It additionally explains an understanding management system and disseminated registering focal points of using cloud computing in knowledge management structures.

3.2 Literature Review

Dave et al. (2013) proposed that the approach of distributed computing has spread out various roads which were not investigated or brindled in the most idealistic fashion. To discover, produce, memory table, talk to or scatter information, the current devices, inventions, or strategies has not held the option to fulfill what a course of action has consistently wanted for its advancement. King and Nabil (2013) suggested that in associations, all things considered live in an undeniably powerful world. A few frameworks exploit this dynamism and progress to new items and plans of action and prosper. Arpaio and Ibrahim (Arpaci 2017) proposed an objective to examine the precursors and outcomes of distributed computing selection in instruction to accomplish information on the board. The discoveries showed that instructive establishments may advance the appropriation of distributed computing in educating by expanding the attention to inform the board reviews. Li et al. (2009) proposed a personal information on the board (PKM) is basic to human getting the hang of, playing, and occupation. Complicating processing, as a rising and noteworthy innovation, is promising to help the individual information to the executives by providing clients with a wide range of versatile administrations. Aksoy and Algawiaz (2014) proposed that the achievement of associations to a great extent relies upon constant interest in determining and securing new information that makes new openings and improves existing performance.

3.3 Knowledge Management and Cloud Computing Benefits and Risks

It is challenging to illustrate the expression of "knowledge," as it has various implications relying upon setting. Knowledge obtaining is the routine of improvement and formation of experiences, achievements, and connections (Li et al. 2009). In addition, cloud can be set apart as open, private, neighborhood, or cross. Customers of on-demand self-organization can course of action circulated registering assets apart from requiring human association, mostly done, but an online self-organization section.

3.3.1 Benefits and Risk

Distributed computing gives a versatile online environmental factor that makes it practical to control a quickened mass of study except for affecting device by and large execution (Mahdi et al. 2020). Keeping the skill organization process, the accompanying necessities can be provided food effectively with distributed computing: Cloud is figuring significantly decreases the science-related expenses. It extends the situation of open-source contributions and shared inclinations all through the globe. It comes down the homegrown methodology and uses related to governance of the framework. The social contraption is made up on recognizable adaptable and trustworthy system.

3.3.2 Risks

Depending on the cloud arrangement utilized (SaaS, PaaS, or IaaS), clients of information administration framework may be incapable to get and survey arrange operations or security occurrence logs (Meneghello et al. 2020). Key events with respect to data life-cycle security in the cloud concerning data the board as-an organization join the accompanying: Data security: Confidentiality, Integrity, Availability, Authenticity, Authorization, Authentication, and Non-renunciation. Data durability or assurance: Techniques for absolutely and effectively discovering data in the cloud, erasing/destroying data, and ensuring the data has been overall removed or rendered unrecoverable must be available and used when required. Data protection and recovery strategies for retrieval and recovery: Data must be open and data support and recovery plans for the huge number must be put up and convincing to hinder information disaster, bothersome data overwrite, and demolition. Data aggregation and deducing: It is phenomenal today to find a customer who doesn't think about Google and the data and information.

3.4 Methods for Security Over Knowledge Management and Cloud Computing

Optimized IT foundation gives quick get for required computing administrations (Nyame et al. 2020). In add-on, giving the proper level of security for knowledge administration organization could be a challenging issue that can be settled by applying cloud computing. Moreover, cloud computing can keep knowledge management up with innovation.

3.4.1 Software Provides Access Control and Identity Management

Identity administration and puzzle to manage are essential capacities required for impenetrable cloud computing. The excellent parent of personality administration is logging on to a laptop framework with a node ID and secret phrase (Song 2020). It genuinely states that an actual persona administration, such every bit is required for cloud computing, requires extra energetic confirmation, authorization, and get to control. Identification and authentication are the cornerstones of most get to manage frameworks. Realization is the act of a patron declaring a character to a framework, lots inner the sample of a username or consumer logon ID to the framework. Authentication is affirmation that the user's claimed persona is secure, and it is frequently achieved through a client's secret phrase at logon.

3.4.2 Passwords

Since passwords can be compromised, they must be ordered. This "one-time catchphrase," or OTP, gives most noteworthy security since a present-day mystery word is needed for each unused logon (Tadejko 2020). A password may well be a course of action of characters that's regularly longer than the disseminated number for a private word. Passwords can be joined by different contraptions, counting tokens, memory cards, and quick cards.

3.4.3 Memory Cards

Memory cards allow a nonvolatile capacity of data, but they don't have any planning capability. A memory card stores mixed passwords and other related recognizing information. A telephone calling card and an ATM card are cases of capacity cards.

3.4.4 Smart Cards

Smart cards grant without a doubt more capability than the memory cards by joining additional dealing with dominance on the cards. These credit-card-size contraptions incorporate chip and capacity and are utilized to store computerized marks, private keys, words, and other private information.

3.4.5   Biometrics

An elective to utilizing passwords for verification in sound or focused get the chance to control is bio-metric. The bio-metric relies upon the Sort 3 confirmation portion (Daud and Rahmann 2017). Bio-estimations are depicted as robotized suggests of seeing or insisting the character of a living individual reliant on physiological or social highlights. In bio-estimations, perceiving affirmation may be a one-to-many of an individual's characteristics of a database of putting missing pictures. Certification may be a reasonable to affirm a title to a character shaped by a man. Bio-metric is utilized for seeing insistence in physical controls and for declaration conflicting controls (Singh et al. 2020). Unlike kinds of bio-metric qualities incorporate facial and palm checks. This degree of bits of knowledge is demanded for professional to-numerous appears in quantifiable limits with exceedingly gigantic databases (Daud and Rahmann 2017). These photographs are amassed amidst the selection strategy and approach to these verbal exchange channels is pressing for the fit development of the bio-metric contraption. Appearing coming up next are ordinary bio-metric homes that are utilized to especially underwrite a person's case: Fingerprints—Fingerprint qualities are acquired and made off. Common CERs are 4–5%. Retina takes a gander at—The eye is set around two deadheads from a digicam and an indistinct light smoothly involves the retina for vein structures. CERs are commonly 1.4%. Iris takes a gander at—A camcorder remotely gets iris models and characters. CER respects are round 0.5%. Hand geometry—Cameras clutch third-dimensional hand qualities. CERs are round 2%. Voice—Sensors trap voice attributes, for instance, throat vibrations and pneumatic power, when the condition talks an explanation. CERs are in the vacillate of 10%. Handwritten imprint factors—The unmistakable characteristics of a man form a fall individual is caught and set down. Fundamental qualities thorough about making weight and pen course. CERs are not, now outflanked out suitable at present. Different systems of bio-metric features join facial and palm checks.

In depth comparison of individual security level, each security type shows their secure ranges. In this secure ranges, biometrics performance is best when compared to other security types. Table 3.1 and Fig. 3.1 show the security types and its individual secure ranges.

In depth comparison of combine security level, each security type combines with another security type and shows their secure ranges. In this combine secure ranges, the biometrics performance is best when compared to other combine security types.

Table 3.1 Security and range

Security type	Individual secure range
ID	20
PIN	40
Memory cards	50
Smart cards	70
Biometrics	90

Biometrics shows the high security level performance. Table 3.2 and Fig. 3.2 show the security types and its combine secure ranges.

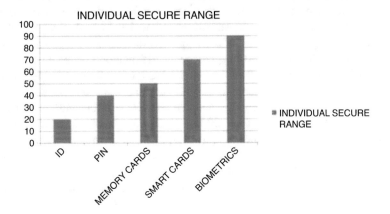

Fig. 3.1 Individual security range

Table 3.2 Security type comparisons

Security type	ID-combine	PIN-combine	Biometrics-combine
ID	20	40	80
PIN	40	25	85
Memory cards	45	50	90
Smart cards	50	65	90
Biometrics	85	90	99

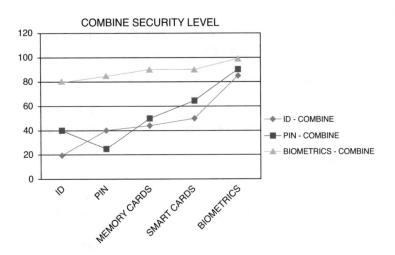

Fig. 3.2 Comparison among security levels

3.5 Autonomic Security

Autonomic computing alludes to a self-managing computing exhibit in which PC frameworks reconfigure themselves in answer to altering prerequisites and are self-healing (Singh et al. 2020). The assurance of autonomic computing will require a few a lengthy time to absolutely materialize, as it have been it provides abilities that can development the safety of information frameworks and cloud computing. The capability of autonomic frameworks to accumulate and get its statistics and recommend or execute preparations can go a long way towards enhancing protection and turning over for restoration from hurtful effects. The autonomic cope with would get logged and matched the statistics and function for an examination to examine the difficulty region. The overseen aspects manage their inward states and have characterized execution traits and associations with different computational factors (Singh et al. 2020). The intention of the supreme self-retrieval method is to maintain the components in operation corresponding to their layout plans.

3.6 Conclusion and Future Work

Working knowledge management techniques should result in improved attaining, or still beyond, business goals. Though knowledge management remains not a constant progression, the thing is dynamic since double features: the enterprise and the knowledge. The carrying out of cloud computing over knowledge management method will power facilitate businesses to achieve rapidity along regarding intelligence expertise and to satisfy corresponding to the constantly shifting constraints of the market. Immediately the cloud computing concept is turning more and more widespread, owing to the immense reduction in the period, price, and attempt for gathering computer software development requirements. It additionally represents a fantastic method for collecting and relocating information. Although, utilizing cloud computing in knowledge management approach has several consequences like threat of documents outflow, IT managerial modifications, and cloud maintenance supplier feasibility. In this article, it reviews the security of integrating confidentiality, belief, and secrecy for knowledge management and cloud computing. Knowledge management desire persist to be important as companies operate simultaneously, distribute data, as well as cooperate with each other on schemes. Holding on the knowledge and actions while distributing and cooperating will be present at most important concern. The forthcoming effort will be a period reduction consequence of applying cloud computing in knowledge management methods of decreasing probability and effect of consequences in cloud computing.

References

M.S. Aksoy, D. Algawiaz, Knowledge management in the cloud: benefits and risks. Int. J. Comput. Appl. Technol. Res. **3**(11), 718–720 (2014)

I. Arpaci, Antecedents and consequences of cloud computing adoption in education to achieve knowledge management. Comput. Hum. Behav. **70**, 382–390 (2017)

N.H.M. Daud, S.A. Rahmann, Advantages of using cloud computing by knowledge management personnel. Int. J. Acad. Res. Bus. Social Sci. **7**(11), 908–916 (2017)

M. Dave, M. Dave, Y.S. Shishodia, Cloud computing and knowledge management as a service: A collaborative approach to harness and manage the plethora of knowledge. BVICAMs Int. J. Inform. Technol. **5**(2), 619–622 (2013)

L. Li, Y. Zheng, F. Zheng, S. Zhong, Cloud computing support for personal knowledge management, in *2009 International Conference on Information Management, Innovation Management and Industrial Engineering*, vol. 4, (IEEE, New York, 2009), pp. 171–174

B. Mahdi et al., Practical approach of knowledge management in medical science. arXiv preprint arXiv:2001.09795 (2020)

J. Meneghello et al., Unlocking social media and user generated content as a data source for knowledge management. Int. J. Knowl. Manag. **16**(1), 101–122 (2020)

G. Nyame et al., An ECDSA approach to access control in knowledge management systems using Blockchain. Information **11**(2), 111 (2020)

A.K. Singh et al., Research and Challenges of Security & Privacy in Internet of Things (IoT), in *2020 International Conference on Computation, Automation and Knowledge Management (ICCAKM)*, (IEEE, New York, 2020)

H. Song, Testing and evaluation system for cloud computing information security products. Proc. Comput. Sci. **166**, 84–87 (2020)

N. Sultan, Knowledge management in the age of cloud computing and Web 2.0: Experiencing the power of disruptive innovations. Int. J. Inform. Manag. **33**(1), 160–165 (2013)

P. Tadejko, Cloud cognitive services based on machine learning methods in architecture of modern knowledge management solutions, in *Data-Centric Business and Applications*, (Springer, Cham, 2020), pp. 169–190

Chapter 4
Exploring Potential of Transfer Deep Learning for Malicious Android Applications Detection

Mohammed Alshehri

4.1 Introduction

Android malware is getting lethal and evasive under the shadow of underlying black investment to fulfill various attackers' motives. In recent years growth in android malware is exponential, which is alarming due to the ubiquitous nature and wealth of information held in android devices. Current malware detection and prevention techniques cannot deliver the required protection to the end-users, which demands a need to find alternative ways to build malware detection and prevention system. Noxious projects or malware are deliberately composed favorable to gram to enjoy different pernicious exercises or cybercrimes. Aggressors need auxiliary and conduct contrasts that cause diverse malware classes, for example, Virus, Worm, Trojan, and Bots. The quick development in the utilization of computerized savvy gadgets, for example, cell phones, smartwatches. The Internet gives a plethora of scope for malware writers to target the android eco-system. In recent years, android malware is being used in various malicious activities like stealing, spying, and financial fraud to country espionage. The growth in the number of android malware is exponential in the last few years. Figure 4.1 shows the growth graph of android malware during 2012–2019.

In general, or especially android malware detection, malware detection is an active research area to prevent malware entry in the end-user's device and keep the user safe from attackers' malicious intention. Malware location is extensively gathered into two classes, which are non-signature-based discovery and mark-based identification. Customarily, discoveries like signature-based (Agarwal et al. 2016)

M. Alshehri (✉)
Department of Information Technology, College of Computer and Information Sciences,
Majmaah University, Majmaah, Saudi Arabia
e-mail: ma.alshehri@mu.edu.sa

© The Author(s), under exclusive license to Springer Nature Switzerland AG 2021
A. Bhardwaj, V. Sapra (eds.), *Security Incidents & Response Against Cyber Attacks*,
EAI/Springer Innovations in Communication and Computing,
https://doi.org/10.1007/978-3-030-69174-5_4

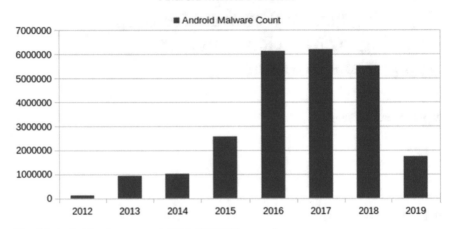

Fig. 4.1 Android malware growth 2012–2019 [AV-test.org]

are utilized regularly in antimalware items, this procedure utilizes a grouping of bytes, which is mark, to identify malware, yet then again, the non-signature-based strategies is an ongoing technique for malware recognition dependent on practices, rule, and various examples of malware. Detection for the signature-based system has high detection accuracy, but these are limited to the signature database only (Allix et al. 2016), requiring regular updates, which is a computationally high task for android devices and limits usability. It also creates the attack time frame, i.e., equal to publicly disclosing malware to signature update. Currently, non-signature-based detection has less accuracy than signature-based techniques, but it eliminates many of the bottlenecks. For the detection of the unknown, zero-day and modern malware, non-sign-based detection techniques are used.

As per news report by Kaspersky (Kaspersky Corporate News 2018), mobile products and technologies detected the following:

- The total repackaged items 3,503,952.
- The Trojan for mobile banking 69,777.
- The new ransomware 68,362.

Attacks on mobile devices' personal data have been increased over the past year by half: from 40,386 unique users in 2018 to 67,500 in 2019. It is not about classic spyware or Trojans, called Stalkerware, and presented in Fig. 4.2 below.

With the expansion in adware dangers in 2019, one design being to individual gather information on cell phones. The measurements show that the quantity of clients assaulted by adware in 2019 is generally unaltered from 2018, as delineated in Fig. 4.3 below.

These markers commonly associate, however not in the situation of adware. we can clarify all these by a few variables, for example establishment bundles of adware produced naturally and can be spread anyplace; however, now and again these bun-

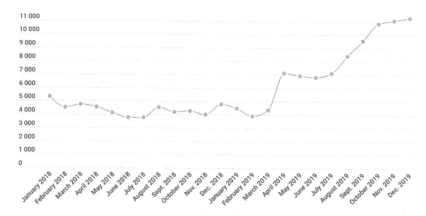

Fig. 4.2 Stalkerware propagation since 2018–2019 (Kaspersky Corporate News 2018)

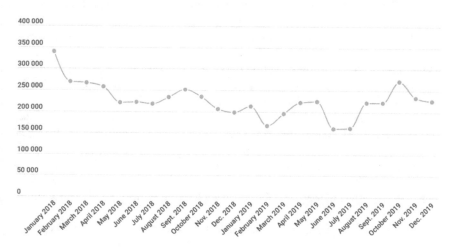

Fig. 4.3 Adware propagation since 2018–2019 (Kaspersky Corporate News 2018)

dles don't arrive at the intended interest group. Sometimes, these packages get detected after the generation and these are not propagated further. Mostly these apps don't contain anything useful, just some of the adware modules (Aung and Zaw 2013). As a result, in 2019, there were 2 trends in a particular stick out, i.e., attaching on user's personal data becomes very often and the detection of trojans.

The non-signature-based detection techniques include data mining approaches, heuristic-based approach, and neural network-based approaches (Conti et al. 2010). These non-signature-based techniques (Darus et al. 2018) utilized malware and benign programs' structural and behavioral features to build a model to classify new files as malware or benign. Permissions, intents, API calls, byte-frequency, etc. are different features (He et al. 2016) which are used to create training and testing sets of android malware classifiers.

In past years, non-signature-based android malware detection mostly involves data mining based techniques such as Decision Tree (DT), Support Vector Machine (SVM), and Naive Bayes (NB) (Idika and Mathur 2007). While heuristic and neural network-based approach were less in research. Heuristic-based approach suffers due to its high false positive, false negative, and expensive training requirement while neural network (Kolosnjaji et al. 2016) based approach was limited in labelled dataset requirements. Recent breakthrough in deep learning (Krizhevsky et al. 2012), i.e., unsupervised learning with a deep network of neural networks (Kumar 2017), has opened a new perspective for non-signature-based malware classifiers. Naeem et al. (2019) explored and experimented the suitability and potential of transfer learning technique, under deep learning for android malware classification with image-based features (Kumar et al. 2016) where the Mal Image (i.e., Image representation of android APK files) were generated from binary form of malware and benign APK samples. Binary to image representation becomes a preprocessing step (Shaha and Pawar 2018) and offers a new secondary level dataset to perform further actions such as extracting image features and training machine learning models (Nataraj et al. 2011). The section II discussed the related work in malware visualization (Pan and Yang 2009) and classification based on image features (Quattoni et al. 2008). Further, section III outlines the proposed work and section IV discusses the experimental setup. Results are presented in section V and section VI lists out the conclusions of the proposed work.

Malware is one of the high positioned danger among top digital dangers of today (Weiss et al. 2016). Aside from conventional mark-based identification strategies, there have been many examinations takes a shot at android malware discovery utilizing AI and information mining methods in the most recent decade. Utilization of deep learning calculations for android malware identification is new heading around there to accomplish higher precision by using the present very good quality equipment abilities.

Although deep learning in malware detection is very new, there have been few works in both commercial and research fronts. Symantec is one of early adopter of deep learning for their commercial product. Symantec's Norton Mobile Security for android uses a deep learning-based back-end detection engine to detect android malwares.

Droid-Sec group which was founded in early 2014 to improve android security has also researched and implemented deep learning to address android malware detection (Yuan et al. 2014). Droid-Sec group has published their research works and has also given an online system based on deep learning-based detection engine to check android app for malware detection. Use of deep learning for malware detection was also demonstrated at Black hat conference by using static features of malware and benign samples. By going through aforementioned progress (Yuan et al. 2016a) of malware detection, it is very easy to understand the power of deep learning for malware detection. Observing the industrial interest in it also suggests the acceptability and use of deep learning-based malware detection. Computer graphic and image processing techniques (Zhao and Qian 2018) have given an alternative new direction to malware research.

4.2 Related Works

While researching the published work on the same topic, the authors identified over 300 publications from IEEE, Elsevier, ACM, and ProQuest from 2013. The authors shortlisted 56 relevant publications using the multi-stage selection process, as illustrated in Fig. 4.4.

Table 4.1 explains and presents the research papers and their selected subcategories as "mobile malware," "image representation," "malware visualization," and "deep learning."

The authors reviewed the closest matching and relevant research papers and industry models, some of which are presented in this section below.

Android is the standard operating system for smartphones. The rapidly rising adoption of android has led to a spike in malware numbers relative to previous years. There are several antimalware security tools intended to defend confidential user data against such attacks on mobile devices. Sabhadiya et al. (2019) analyzed android malware, and its approaches focused on the in-depth analysis used to assault malware and antivirus programs for android systems. The writers explored numerous techniques such as Maldozer, DroidDetector, DroidDeepLearner, DeepFlow, Droid Dover, and Droid Deep. The developers have applied a deeply learned model to determine whether malware is corrupted or not without installation instantly.

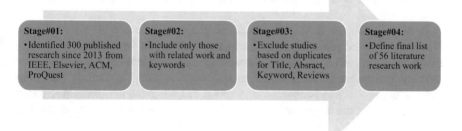

Fig. 4.4 Multi-stage literature selection

Table 4.1 Android malware literature findings

Paper classifications	Stage 1	Stage 2	Stage 3	Stage 4	Final count
Mobile malware	96	75	45	25	18
Image representation	63	49	29	16	12
Malware visualization	87	68	41	22	16
Deep learning	54	42	25	14	10
	300				56

Android malware is prevalent these days because reliable sources do not build apps. People insert their details, saving cards, and more, think these apps can keep them active or allow them to recall those crucial works that we seem to miss during this busy life routine. In these instances, it would be very beneficial to detect vulnerabilities even before launching a program. It could also deter cybercrimes. HR (2019) suggested in this paper the need for a fully connected deep learning model in the detection of android malware. The key characteristically of the work being suggested is android's malware identification right before download, android malware name, proved a handy version kit of approx. 94.65%. This model also knows all the characteristics of all design combinations. It requires rigorous analysis and measurement to achieve a high degree of precision.

As smartphones worldwide become saturated, mobile malware is beginning to expand in sophistication and amount, subverting the biggest target set with the most precious benefits. Around the same time, profound learning is becoming an effective data processing and modelling technique that demonstrates considerable improvement with managed learning when identifying new and unexpected malicious apps. Despite this development, it should further enhance the appropriate use of deep information in malware identification through additional data and model creation. In this study, McGiff et al. (2019) considered android malware detection of numerous extractable data groups using statics analysis techniques. In specific, authorization and hardware data in diverse multimodal feedback scenarios and in deep network formats are used. The authors showed by experimental study that the mixture of both data sets could boost their overall efficiency, achieving 94.5% accuracy. Furthermore, it was found with a narrow grid search that the largest multi-module network takes the lowest training time and greater precision than our other models.

In recent years, the security issues for android smartphone users have suffered. The protection and privacy of the smartphone malware in the network is severely endangered. Li et al. (2018a) used this paper to identify android malware using deep learning approaches to build an automatic detection motor to identify malicious apps' families. The assessment findings reveal that while the fine-grained malware families identify, the engine will identify 97% of malware at a 0.1% false-positive rate (FPR).

Android smartphones accounted for around 87% of the mobile industry up to 2017. The big market even promotes android malware growth. Currently, more than 38,000 are found every day for malware targeting android devices. Due to exponential development in smartphone applications and anti-reverse technologies, both forms of malware are becoming more difficult to classify. Xiao (2019) suggested an approach to detection that specifically apprises malware from Dalvik bytecodes based on the deep learning technique (CNN) to solve problems in current detection strategies like data obfuscation and restricted coding scope. Our model is averaging 0.22 s detection, which is much less than most current models of detection. In the meantime, our model is overall more than 93% reliable.

Mobile apps are growing in every life area, including banking and finance, in today's busy world. Cybercrimes take the form of hacking and ransomware as a motivation. All cannot know what permissions to accept and what to reject when downloading the smartphone applications. Where malware and malicious apk files begin to allow all permissions, some of them can access the mobile device. Many machine learning approaches were implemented to solve this problem, but in real-time implementations, they were not considerably accurate. Sirisha et al. (2019) based on the detection of malware that could reach android through the use of a deep neural network model through permissions. In real-time android apk files, the suggested solution detects the malware powered by authorization with greater than 85% accuracy.

In everyday life, smartphones and computer devices easily become indispensable. Since 2012, Android has become the most popular smartphone OS. Android's open existence; however, countless malware is concealed in various innocuous android applications, which significantly jeopardizes android protection. DL is a new field of MLresearch that has been devoted more and more to artificial intelligence. In the research, Yuan et al. (2016b) suggested that static analysis functions be combined with dynamic analyses of android applications and that deep learning method be used to classify malware. We develop an android malware detection engine for online learning to automatically identify whether an app is a malware. We extensively checked and evaluated the functionality that deep learning utilizes to identify malware across thousands of android applications. The findings indicate that DL is accurate for android malware characterization, incredibly successful when additional training data is available. The accuracy of detection can be 96.76%, which exceeds conventional ML techniques. An examination of 10 standard antivirus tools indicates the need to develop our malware detection capability.

The increasing amount of android malware and its variety have considerably undermined the efficiency of traditional security mechanisms and therefore, detecting android adware is often shielded away from fresh and unknown malware. The authors proposed DroidDeep, a malware identification solution for the android platform based on the profound learning paradigm, to overcome these limitations. The new field of computer study that draws growing interest to artificial intelligence is deep learning. To do this, Su et al. (2016) have first derived five forms of features from android static analysis. The authors create a profound learning paradigm for android applications. Finally, unknown android malware can be identified using the features that are taught. DroidDeep executes multiple current malware identification protocols and achieves 99.4% precision in an experiment along 3986 benign software and 3986 malware. Besides, DroidDeep can achieve an outstanding runtime efficiency that always makes it suitable to adapt to android malware's larger-scale real-world detections.

In recent times, smartphones are a crucial part of our everyday lives. With a larger number of android applications (apps) built to make user experience more comfortable with number of features, android-based smartphones have become the most common alternative. Li et al. (2018b) suggested an android malware characterization and recognition method that mostly uses a DL algorithm to resolve the critical need for detection of malware to cope with malicious applications (malware), which seriously endanger android smartphones' protection. The method achieves over 90% accuracy with just 237 elements and demonstrates detailed experimental performance.

With the immense development of the Internet sector, the mobile Internet was effectively integrated in daily practice. Yet android has several significant security concerns. Mu et al. (2019) have been conducting research to identify android malware based on deep learning text ranking technologies. Cuckoo's android malware series is retrieved and used by the developers to address the identification problem using text processing techniques. To test the results, the authors compared Dalvik with the bi-LSTM. Cuckoo is above Dalvik and is 96.74% correct for the exactness of API extraction. The author uses Cuckoo Sandbox as an API removal system to compare it with GRU, BGRU, and LSTM to check the different models' results further. The findings demonstrate Bi-LSTM's highest accuracy.

The versatility and extensibility of android have developed it into a popular mobile network and a pleasant internet for a driver. Because of its popularity and usability, this is the main subject of malware. Malware represents a significant threat to consumer safety, money, and hardware and file storage. Zhu et al. (2020) developed a new SEDMDroid malware detection application by examining the malware behavior. This deep learning process stacking group is a two-tier framework, with an ensemble of simple MLP classifiers and a Vector Machine (SVM) classification. This is a fusion support group. These are two-pronged disturbances strategic to ensure the consistency and variety of the specific classifiers by sample disturbances and usable space and to take advantage of the eligible base classifiers' effects to optimize classification. For the testing the proposed method, multilevel static data collecting with an average precision of 89.29%, covering entry, responsive APIs, control mechanisms, and entry rates is used.

With the number, variance, and speed dimensions of significant data growing, data owners and researchers did not quickly perceive other dimensions, such as truthfulness, meaning, variability, and place. The idea that the data can be used in computer analysis, such as classification or grouping, still doesn't seem apparent. Canbek et al. (2018) suggested four methods of quantizing parameters to systematically profile and see strong and weak features of data sets obtained from various resources. The technique recommended in the writing and security industry, to be specific Android Malware Project, Drebin, Android Malware Dataset, Android Botnet, and Virus Complete, 2018, is obvious in five android versatile malware datasets. The findings indicate that the current profiling approaches offer a comparative perspective into datasets and that researchers are guided to obtain wider yet more noticeable, better qualitative, and more comprehensive datasets.

The distribution of android apps is increasing as malware writers use malware vulnerabilities to access personal or confidential information with poor intentions that are mostly of economic significance, with smartphone users experiencing malicious applications (malware). Wang et al. (2016) used different techniques and researched malware identification to protect network stability and retain user trust. As malware becomes more likely to hide its malicious intent by using code obstructiveness, it is essential to sustain the rate of malware improvement for malware detection techniques.

At the moment, most current android application malware identification methods use simultaneous pattern matching, which is exceptionally successful but restricted to what computers have seen before. However, when it comes to detecting malicious software, their output degrades considerably. The authors suggested a Malware classification and recognition method that uses a profound learning algorithm to solve the ongoing need to identify malware in order for the issue to be more autonomous. Test findings suggest that, as opposed to current commonly used malware detection methods, the DroidDeepLearner approach achieves good efficiency. The widespread use of smartphones has dramatically increased the amount of malware. Among intelligent devices, android is the most common malware system because of Kim et al. (2019), new architecture is introduced for the identification of android malware. The system uses a number of features to represent the properties of android applications from different angles to refine their features by using our current or similarity-based approach to remove malware effectively. It is also suggested that a multimodal system of profound learning be needed as a malware model. Investigation is the first multimodal analysis to use in the identification of android malware. In consideration to detection model, the advantages of different function forms could be maximized. We performed numerous experiments with 41,260 samples to test our results. The performance of our model has been compared with other applications of deep neural networks. We have tested our system in numerous ways, including productivity in design changes, the utility of various features, and our view technique. Furthermore, we correlated our system's efficiency with that of other approaches, including profound learning methods.

Android malware identification relies on the static and dynamic functional vector extraction of android apps. Static analysis provides again over dynamic analysis because it contains both byte and manifest files and the authorization to use the source code, while dynamic analysis of the APK files involves such features as a device call number, a network url, etc. Change variable in the dataset attributable to product version changes provides a problem for the current method to categorize the product as malicious or benign. Kaushik and Yadav 2018) developed a neural network combination of automatic resources that gathers and updates functionality vectors. The neural network uses this dataset to enhance classification and classifications of malicious, friendly and unstable domain classes for its reinforcement training. Tensor flow is used to create a neural network that learns from information derived in a functional vector and categorizes applications as malicious, benevolent, or unable to tell. For a data set of thousand sample applications with over 15 separate feature vectors derived from designated automatic feature selection modules, over 80% accuracy has been achieved.

Android has achieved tremendous success of intelligent users over the last few years as the most powerful OS. As Android OS is popular and available, it is the tempting target of malicious apps that can seriously threaten financial institutions, companies, and people with protection. Traditional antimalware programs are not enough to battle sophisticated malware that has just been developed. Automatic malware identification systems are also urgently required in order to mitigate the likelihood of malignancies. Nowadays the machine learning-based technique shows impressive results in malware analysis where most approaches are low-level students like Logistic Regression (LR). The paper proposes a profound learning system for recognition of malware for Masum and Shahriar (2019). However, the approach introduced is a deep learner who beats sophisticated conventional techniques in machine learning. To test the system, the authors conducted all experiments on two datasets (Malgenome-215 & Drebin-215) of android applications. The test results illustrate the solidity and reliability of the system.

As an efficient approach to collecting longer-term, time dependencies at arbitrary duration series, Long-Term Repeating Speed (LSTM-RNN) has been observed. In a larger number of android program package, i.e., APK files of the Cyber Security Data Mining Competitors (CDMC 2016), Vinayakumar et al. (2017) posed a wide variety of android permissions, including permissions from regular, hazardous, and signature or device groups, and the task of classification of android malware. By repeating LSTM layer through word bag embedding the series of android permissions, the extracted features are fed to a dense layer with a triggering method that is nonlinear, including sigmoid for the classification. We have performed multiple experiments with various network parameters and network configurations to determine the ideal network configuration and parameters selectively. Both tests have a learning rate in the range [0.01–0.5] of up to 1000 epochs. In contrast to the recurrent neural network (RNN), all of the LSTM network architecture have performed well in the fivefold cross valid classification parameters. Moreover, the LSTM test range, supplied by CDMC2016, has achieved the highest accuracy in the real world, namely 0.897. This is partly because LSTM has a vast and complex memory processing unit, enabling you to understand the temporal actions with sparse android sequence representations quickly. Therefore, we claim that it is more fitting to extend the LSTM network to the classifications of android malware.

With its transparency and versatility characteristics, android has become the most popular smartphone app. But it has also been mobile malware's most controlled device. The users need a swift and accurate method of detection. This article suggests a two-layer malware detection system for android applications, introduced by Feng et al. (2020). The first layer is a static malware detection model based on authorization, intent and component information. The detection rate of the first layer, 95.22%, incorporates the static and the dynamically linked neural network to detect the malware and assesses their efficacy by trial. Input is then rendered into the second layer by the output (benign applications of the first layer). In the second layer, a new CACNN approach is used to detect malware through the network traffic functionality of applications, cascading both CNN and AutoEncoder. The identification rate for the second class (2-classifier) is 99.3%. In addition, malware by class (4-classifier) and malicious (40-classifier) families can be found in the current two-layer model. The rates of identification are 98.2% and 71.48%, respectively. The experimental findings suggest that our two-stage approach can accomplish semi-supervised learning and boost malicious android APP detection effectively.

Zhang et al. (2018) suggested a deep learning-based android malware detection system. The input data, the static, and the dynamic features are divided into two sections. In this article, the android code is carefully evaluated to obtain

485-dimensional static characteristics. Via simulator operation, the dynamic time characteristics of the alteration are achieved and the 23-dimensional dynamical characteristics of the neural RNN phase are achieved. The mixture of static and dynamic properties is a unique way to detect android malware from our experiments. This profound learning model will considerably increase detection efficiency compared to standard low-speed models as LR and SVM, based on a comprehensive free android dataset. This way, the protection of the android device is very adequate and functional (Table 4.2).

Few other research papers which are relevant and close to our research are presented below.

In an early work of binary visualization, Shaid and Maarof (2014) presented a visual study of binary fragments and give extensible and visual taxonomy of these primitive binary fragments. In this work, five main media types, i.e., text, image, audio, video, and application, and other classes such as random and repeated binary data, are listed with their subcategories and description. The authors used malware image (converting binary malware samples to greyscale image) to classify them in 25 distinct malware families. The authors have also compared their result with traditional static and dynamic methods and claim an accuracy of 98% with 40 times low computation cost. Image-based classification is robust against packed and obfuscated malware and gives the same result while traditional methods fail against such samples.

In another exploration work, Sun and Qian (2018) proposed a malware examination technique that utilizations imagined pictures and entropy diagrams to distinguish and group new malware and malware variations. The authors presented a detailed experimental result of two datasets; one has malware and benign samples, and others having only malware families. As future work, authors have suggested a dynamic analysis-based instruction-level analysis for packed malware. In recent years, many research works have applied deep learning for android malware detection and specially used image representation of android samples. The authors used deep learning for malware classification. In this work, authors have used recurrent neural networks (RNN) and convolutional neural networks (CNN) for building malware families classifiers. With the minhas features generation method, the authors achieved a good result for the classification. Tuvell et al. (2012) used disassembly representation of android files to generate the image and used this image dataset to build an android malware classifier. In very recent work, Naeem et al. have used image representation and SIFT and GIST features for building a cross-platform malware classification system.

Based on the available literature, it is observed that deep learning-based malware detection is an active area of research, but there are very little works for android malware detection. In this proposed work, the potential of transfer learning is explored for android malware classification.

Table 4.2 Matching references and keywords

References	Matching references and keywords
Sabhadiya et al. (2019)	The android operating system, invasive software, artificial intelligence, mobile computing, smartphones, Android malware detection, standard smartphone operating system, rapidly growing acceptance, antimalware programs, mobile user systems, antivirus programs, detection techniques, deep learning, Android application, Droid Deep Learner, Deep Flow, Droid Delver, Malware, Deep learning, Feature extraction, Detectors, Task analysis, Biological neural networks, Android Security, Malware Detection Technique, Deep Learning-based Malware Detection
HR (2019)	Smart phones, Full connected DL model, static analysis, android malware detection, Feature extraction, Machine learning, Android Malware Classification, ML, Permissions, APKs, i.e., application packages, DL, Data Mining, Data Extraction, Preprocessing, Vector Representation, Behavioral Analysis, Deep Learning Dense Model, Random Forest Classifier, Virus
McGiff et al. (2019)	Android OS, Computational complexity, Data analysis, Invasive software, AI, Mobile computing, Program diagnostics, Smart phones, Deep learning, Data analysis, Unforeseen malicious, Supervised learning, Model construction, Static analysis techniques, Multiple extractable data classes, Multimodal input scenarios, Deep network shapes, Experimental analysis, Largest multimodal network, Towards multimodal learning, Android malware detection, Complete saturation Mobile malware, Valuable rewards, Hardware, Malware, Feature extraction, Smart phones, Deep learning, Data mining, Neural networks, Deep Learning, Malware Detection, Performance Tuning
Li et al. (2018a)	Android, Data privacy, Feature extraction, Invasive software, AI, Mobile computing, Android malware detection, Deep learning, Network security, Mobile malware, Network privacy, Automatic detection, Android smartphone, Malware, Biological neural networks, Androids, Humanoid robots, Security, Neurons, Android Malware Detection, Deep Neural Network, Fine-grained Classification, Smartphones Security
Xiao (2019)	Convolutional neural net, Android operating system, Artificial intelligence, Mobile computing, Smart phones, CNN-based Android malware detection, Android devices, Android smartphones, Mobile application program, Anti-reverse-engineering technique, Deep learning technique, Image-inspired Android malware detection, Dalvik bytecode, Malware, Feature extraction, Deep learning, Transforms, Optimization, Android Malware Detection, Deep learning
Sirisha et al. (2019)	Android, Crime, Invasive software, Artificial intelligence, Mobile computing, Neural nets, Smart phone, Deep neural network model, Permission driven Malware, Time android, Deep learning techniques, Busy world, Mobile application, Finance, Mobile apps, Malicious apk files, Real-time applications, Banking cybercrimes, Malware, Mathematical model, Biological neural networks, Machine Learning, Sequential neural network
Su et al. (2016)	Invasive software, AI, Machine learning, Mobile computing, Program diagnostics, Android apps, Static analysis, Deep learning model, Droid-Deep, Android malware, Feature extraction, Androids, Humanoid robots, Smart phones, Mobile communication, Android malware, Notebook computers, Deep learning, Security, Static analysis

(continued)

Table 4.2 (continued)

References	Matching references and keywords
Mu et al. (2019)	Application program interface, Internet attacks, Mobile computing, Pattern classification, Recurrent neural nets, Text analysis, Mobile Internet, Security issues, Text classification technology, Deep learning, Android malware, API sequence, API extraction method, Cuckoo sandbox, Android malware detection, API calls, Internet industry, Android, Malware classification, Bi-LSTM
Canbek et al. (2018)	Android, Big Data, Mobile computing, Mobile malware datasets, Internalized datasets, Profiling methods, Botnet Dataset, Malware Dataset, Android Malware Genome Project, Machine learning, Data owners, Big datasets, Mobile applications, Genomics, Bioinformatics, Aerospace electronics, Data profiling, Data quality, Feature engineering
Kaushik and Yadav (2018)	Android, Feature extraction, Invasive, Program diagnostics, Android application, Deep learning, Android malware, Feature vector extraction, Static analysis, Dynamic analysis, Manifest files, APK files, Feature vector updating, Neural network, Network url, Reinforcement training, Tensor flow, Feature vector collection, Biological neural networks, Malware Tools, Recurrent neural networks, Infected apk, Deep learning, Feature detection, Malicious apk, Feature vector
Masum and Shahriar (2019)	Pattern classification, Regression analysis, Android operating system, Malicious apps, Security threat, Automatic malware detection, Malicious activities, Machine learning algorithms, Malware classification, Cutting-edge machine, Android apps, Deep learning, Neural network, Android malware detection, Humanoid robots, Smart phones, Support vector machines, Biological neural networks, Neurons, Android malware, Android security
Vinayakumar et al. (2017)	Android Data mining, Pattern classification, Recurrent neural, Deep android malware detection, Short-term memory recurrent, Neural network, LSTM-RNN, Long-range temporal dependencies, Android application, Cyber Security, Data Mining Competition, Recurrent LSTM layer, Bag-of-words embedding, Dense activation layer, Nonlinear activation function, Optimal parameters, Network structure, Network configurations, Classification settings, fivefold cross validation, CDMC2016, Complex memory processing, Android permissions sequences, Permission based Android malware classification, Android malware classification, Network parameters, Android malware dataset, Machine learning, Recurrent neural networks, Feature extraction, Malware detection and classification, Permissions APK, Recurrent neural network (RNN), Long short-term memory (LSTM)
Feng et al. (2020)	Mobile Android, CNN, Invasive software, AI, Mobile computing, Program diagnostics, Telecommunication traffic, Static malware detection, Full connected neural network, Malicious Android applications, i.e., apps, Android malware detection, Mobile platform, Mobile malware, CACNN, Two-layer DL(deep learning), Auto Encoder, Network traffic, Malware, Feature extraction, Smartphones, Deep learning, Mobile applications, Malware detection, Deep learning, Network traffic

4.3   Transfer Learning for Android Malware Detection

Transfer learning, in general, is utilizing pre-learned knowledge to a different task than the earlier task. Deep learning works in a very similar way where a neural network trained on a problem can be saved and further fine-tuned with other datasets.

This proposed work uses image processing techniques and transfer learning for the classification of android malware. To use techniques from image processing for android malware classification, the dataset must be invalid image format, so first malware and benign samples were converted into image and then deep learning algorithms were used to build classifiers.

- *Dataset Preparation*
  For the experiment and building the proposed system, the benign and malicious android samples downloaded from AndroZoo project an extensive collection of android APK files with benign and malicious samples. After downloading the android samples, each sample was processed for duplicate removal, and the class label, i.e., malware or benign, was verified based on VirusTotal results. The adequately filtered and the grouped sample was converted to RGB image by reading byte-byte and giving them value accordingly. At the end of the dataset preparation step, there were 2000 images of malware and benign samples, respectively, and were inputted to the feature extraction modules.
- *Feature extraction*
  In this proposed android malware classification task, the dataset has two classes (malware and clean) so setup is created to train the model with pre-trained models. For example, the ResNet-18 (trained on ImageNet Data) is used by changing the last layer from (input feature = 512, output class = 1000) to (input feature = 512, class = 2) and trained again for fine-tuning. Similarly, other pre-trained models such as AlexNet, Vgg16, GoogleNet, and MobileNet same configuration are carried out.

4.4   Experimental Setup

This chapter proposes android malware classification using image representation of android files and transfers learning using pre-trained image-based deep learning models. The proposed method has four steps, as shown in Fig. 4.5.

In step 1, which is common to other methods also, the malware and benign samples are collected and pre-processed. Preprocessing consists of duplicate removal and labels the sample with the correct class label, i.e., (malware or benign). Duplicate removal is performed by calculating & MD5 hash of each sample. Randomly only one sample is retained comparing in the dataset if the MD5 hash match with others. Labeling of the sample is performed by using VirusTotal API services. With the cleaned dataset of step 1, each android APK sample was converted to an RGB image in step 2. According to RGB format, the conversion module takes APK files as input and after reading each byte and converting them into pixel values, output an RGB image file as shown in Fig. 4.6.

After converting the whole android dataset into image format, each image's texture features were extracted in step 3 using pre-trained models. The feature vector of step 3 is Input to different classification module of step 4, step 4 is the output of

Fig. 4.5 Transfer
learning-based android
malware classification

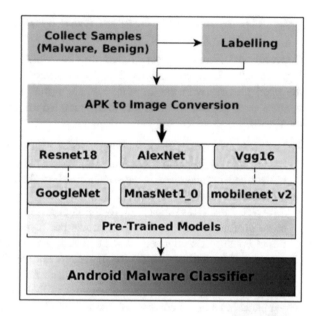

Fig. 4.6 Left: Symbolized
APK file, Right:
Equivalent RGB file

different pre-trained models, and the android malware classifier is the output of this
final step. The purpose of this paper is to explore the potential of transfer learning
to build an android malware classification system that can classify a given new
android APK file into malware or benign class. In other section IV-A, few popular
pre-trained image models are explained.

4.4.1 Deep Learning: Convolutional Neural Network

Convolutional Neural Network (CNN, or ConvNet) is an alternate sort of multi-layering network which are worked for the recognition of visual examples got from the pixel pictures with the assistance of negligible pre-preparing. The ImageNet venture is a picture information base intended to use in object acknowledgment work.

- LeNet-5: LeNet-5, a spearheading 7-level CN (convolutional organization) by LeCun et al. in 1998, arranges digits applied by the quantity of banks to perceive transcribed numbers on (checks) digitized in 32×32 pixel greyscale input pictures.
- AlexNet: AlexNet is fundamentally the same as engineering as LeNet however is more profound, with numerous channels per layer with stacked CN (convolutional layers). It comprised 11×11, 5×5, 3×3, convolutions, max pooling, dropout, information growth, ReLU enactments, SGD with force. It appended ReLU enactments after each convolutional and full associated layer.
- GoogleNet/Inception: GoogleNet network utilized a CNN propelled by LeNet yet executed a novel component that is named a commencement module. It utilized cluster standardization, picture twists, and RMSprop. GoogleNet depends on a few little convolutions to decrease the quantity of boundaries definitely.
- VGGNet: VGGNet comprises of 16 convolutional layers and is extremely engaging a direct result of its uniform design. It has just 3×3 convolutions, however heaps of channels and comprises of 138 million boundaries.
- ResNet: Residual Neural Network (ResNet) is a novel engineering with skip associations and highlights monstrous group standardization. These skip associations are otherwise called gated units or gated intermittent units.

Table 4.3 summarizes the performance of these pre-trained models based on the online image classification competition. This is used in this work to provide a comparative understanding of each of these models.

All the experiments are carried out on the Google cloud platform using the Colab Research environment. This platform helped us to use GPU and fulfilled the computation requirements.

Table 4.3 Classifiers result for performance matric

Year	CNN literature	Rank	Error rate %
1998	LeNet	n/a	n/a
2012	AlexNet	First	15.3
2013	ZFNet	First	14.8
2014	GoogLeNet	First	6.67
2015	VGGNet	Second	7.3
2016	ResNet	First	3.6

Table 4.4 Model performance

Classifiers	Accuracy
Resnet18	77
AlexNet	80
SqueezeNet 110	75
Vgg16	74
DenseNet	70
MobileNet	73
GoogleNet	79

4.5 Experimental Setup

Each of the pre-trained models fine-tuned with training, and 500 Epoc was used. The result of these models measures based on accuracy as per equation 1 and is mentioned in Table 4.4. Table 4.4 shows that each model's performance is above the baseline classifiers but not up to the acceptable accuracy. With the convent feature extractor, we can achieve better results for similar tasks.

4.6 Conclusion

Transfer learning is innovative because it saves time and the computational requirement of deep learning model training, a resource and time-consuming. The proposed work's primary focus is to explore the potential of transfer learning for android malware classification. It has been observed that the model trained for object classification is not very suitable for malware classification. Such cross-domain transfer learning is novel, and so it requires further in-depth research. The low accuracy can be addressed using a model trained for malware classification and again fine-tuned with new samples. One of the other reasons for low accuracy could be attributed to fewer samples in the dataset due to computation limitations. In further research, the subject can be re-explored with more samples and a different trained model, especially for binary classification.

References

R. Agarwal, P.K. Singh, N. Jyoti, H.R. Vishwanath, P.R. Prashanth, System and method for non-signature based detection of malicious processes, US Patent 9,323,928, 26 Apr 2016

K. Allix, T.F. Bissyand'e, J. Klein, Y. Le Traon, Androzoo: Collecting millions of android apps for the research community, in *2016 IEEE/ACM 13th Working Conference on Mining Software Repositories (MSR)*, (IEEE, New York, 2016), pp. 468–471

Z. Aung, W. Zaw, Permission-based android malware detection. Int. J. Sci. Technol. Res. **2**(3), 228–234 (2013)

G. Canbek, S. Sagiroglu, T. Taskaya Temizel, New techniques in profiling big datasets for machine learning with a concise review of android mobile malware datasets, in *2018 International Congress on Big Data, Deep Learning and Fighting Cyber Terrorism (IBIGDELFT), Ankara, Turkey*, (2018), pp. 117–121. https://doi.org/10.1109/IBIGDELFT.2018.8625275

G. Conti, S. Bratus, A. Shubina, A. Lichtenberg, R. Ragsdale, R. Perez-Alemany, B. Sangster, M. Supan, A visual study of primitive binary fragment types, in *White Paper, Black Hat USA*, (2010)

F.M. Darus, S.N.A. Ahmad, A.F.M. Ariffin, Android malware detection using machine learning on image patterns, in *2018 Cyber Resilience Conference (CRC)*, (IEEE, New York, 2018), pp. 1–2

J. Feng, L. Shen, Z. Chen, Y. Wang, H. Li, A two-layer deep learning method for android malware detection using network traffic. IEEE Access **8**, 125786–125796 (2020). https://doi.org/10.1109/ACCESS.2020.3008081

K. He, X. Zhang, S. Ren, J. Sun, Deep residual learning for image recognition, in *Proceedings of the IEEE conference on computer vision and pattern recognition*, (2016), pp. 770–778

S. HR, Static analysis of android malware detection using deep learning, in *2019 International Conference on Intelligent Computing and Control Systems (ICCS), Madurai, India*, (2019), pp. 841–845. https://doi.org/10.1109/ICCS45141.2019.9065765

N. Idika, A.P. Mathur, *A survey of malware detection techniques*, vol 48 (Purdue University, West Lafayette, IN, 2007)

Kaspersky Corporate News (2018), https://www.kaspersky.com/about/press-releases/2019_the-number-of-mobile-malware-attacks-doubles-in-2018-as-cybercriminals-sharpen-their-distribution-strategies. Accessed 25 Sept 2020

P. Kaushik, P.K. Yadav, A novel approach for detecting malware in android applications using deep learning, in *2018 Eleventh International Conference on Contemporary Computing (IC3), Noida*, (2018), pp. 1–4. https://doi.org/10.1109/IC3.2018.8530668

T. Kim, B. Kang, M. Rho, S. Sezer, E.G. Im, A multimodal deep learning method for android malware detection using various features. IEEE Trans. Inform. Forens. Secur. **14**(3), 773–788 (2019). https://doi.org/10.1109/TIFS.2018.2866319

B. Kolosnjaji, A. Zarras, G. Webster, C. Eckert, Deep learning for classification of malware system call sequences, in *Australasian Joint Conference on Artificial Intelligence*, (Springer, Berlin, 2016), pp. 137–149

A. Krizhevsky, I. Sutskever, G.E. Hinton, Imagenet classification with deep convolutional neural networks, in *Advances in Neural Information Processing Systems*, (2012), pp. 1097–1105

A. Kumar, A framework for malware detection with static features using machine learning algorithms. PhD thesis, Department of Computer Science, Pondicherry University, 2017

A. Kumar, K.P. Sagar, K.S. Kuppusamy, G. Aghila, Machine learning based malware classification for android applications using multimodal image representations, in *2016 10th International Conference on Intelligent Systems and Control (ISCO)*, (IEEE, New York, 2016), pp. 1–6

D. Li, Z. Wang, Y. Xue, Fine-grained android malware detection based on deep learning, in *2018 IEEE Conference on Communications and Network Security (CNS), Beijing*, (2018a), pp. 1–2. https://doi.org/10.1109/CNS.2018.8433204

W. Li, Z. Wang, J. Cai, S. Cheng, An Android malware detection approach using weight-adjusted deep learning, in *2018 International Conference on Computing, Networking and Communications (ICNC), Maui, HI*, (2018b), pp. 437–441. https://doi.org/10.1109/ICCNC.2018.8390391

M. Masum, H. Shahriar, Droid-NNet: Deep learning neural network for android malware detection, in *2019 IEEE International Conference on Big Data (Big Data), Los Angeles, CA, USA*, (2019), pp. 5789–5793. https://doi.org/10.1109/BigData47090.2019.9006053

J. McGiff, W.G. Hatcher, J. Nguyen, W. Yu, E. Blasch, C. Lu, Towards Multimodal Learning for Android Malware Detection, in *2019 International Conference on Computing, Networking and Communications (ICNC), Honolulu, HI, USA*, (2019), pp. 432–436. https://doi.org/10.1109/ICCNC.2019.8685502

T. Mu, H. Chen, J. Du, A. Xu, An Android malware detection method using deep learning based on API calls, in *2019 IEEE 3rd Advanced Information Management, Communicates, Electronic and Automation Control Conference (IMCEC), Chongqing, China*, (2019), pp. 2001–2004. https://doi.org/10.1109/IMCEC46724.2019.8983860

H. Naeem, B. Guo, F. Ullah, M.R. Naeem, A cross-platform malware variant classification based on image representation. KSII Trans. Internet Inform. Syst. **13**(7), 3756–3777 (2019)

L. Nataraj, S. Karthikeyan, G. Jacob, B.S. Manjunath, Malware images: Visualization and automatic classification, in *Proceedings of the 8th International Symposium on Visualization for Cybersecurity*, (ACM, New York, 2011), p. 4

S.J. Pan, Q. Yang, A survey on transfer learning. IEEE Trans. Knowl. Data Eng. **22**(10), 1345–1359 (2009)

A. Quattoni, M. Collins, T. Darrell, Transfer learning for image classification with sparse prototype representations, in *2008 IEEE Conference on Computer Vision and Pattern Recognition*, (IEEE, New York, 2008), pp. 1–8

S. Sabhadiya, J. Barad, J. Gheewala, Android Malware Detection using Deep Learning, in *2019 3rd International Conference on Trends in Electronics and Informatics (ICOEI), Tirunelveli, India*, (2019), pp. 1254–1260. https://doi.org/10.1109/ICOEI.2019.8862633

M. Shaha, M. Pawar, Transfer learning for image classification, in *2018 Second International Conference on Electronics, Communication and Aerospace Technology (ICECA)*, (IEEE, New York, 2018), pp. 656–660

S.Z.M. Shaid, M.A. Maarof, Malware behavior image for malware variant identification, in *2014 International Symposium on Biometrics and Security Technologies (ISBAST)*, (IEEE, New York, 2014), pp. 238–243

P. Sirisha, B.K. Priya, K.A. Kunal, T. Anuradha, Detection of permission driven malware in android using deep learning techniques, in *2019 3rd International conference on Electronics, Communication and Aerospace Technology (ICECA), Coimbatore, India*, (2019), pp. 941–945. https://doi.org/10.1109/ICECA.2019.8821811

X. Su, D. Zhang, W. Li, K. Zhao, A deep learning approach to android malware feature learning and detection, in *2016 IEEE Trustcom/BigDataSE/ISPA, Tianjin*, (2016), pp. 244–251. https://doi.org/10.1109/TrustCom.2016.0070

G. Sun, Q. Qian, Deep learning and visualization for identifying malware families, in *IEEE Transactions on Dependable and Secure Computing*, (2018)

G. Tuvell, D. Venugopal, M. Pfefferle, Non-signature malware detection system and method for mobile platforms, US Patent 8,312,545, 13 Nov 2012

R. Vinayakumar, K.P. Soman, P. Poornachandran, Deep android malware detection and classification, in *2017 International Conference on Advances in Computing, Communications and Informatics (ICACCI), Udupi*, (2017), pp. 1677–1683. https://doi.org/10.1109/ICACCI.2017.8126084

Z. Wang, J. Cai, S. Cheng, W. Li, DroidDeepLearner: Identifying Android malware using deep learning, in *2016 IEEE 37th Sarnoff Symposium, Newark, NJ*, (2016), pp. 160–165. https://doi.org/10.1109/SARNOF.2016.7846747

K. Weiss, T.M. Khoshgoftaar, D.D. Wang, A survey of transfer learning. J. Big Data **3**(1), 9 (2016)

X. Xiao, An image-inspired and CNN-based android malware detection approach, in *2019 34th IEEE/ACM International Conference on Automated Software Engineering (ASE), San Diego, CA, USA*, (2019), pp. 1259–1261. https://doi.org/10.1109/ASE.2019.00155

Z. Yuan, Y. Lu, Z. Wang, Y. Xue, Droidsec: deep learning in android malware detection, in *ACM SIGCOMM Computer Communication Review*, vol. 44, (ACM, New York, 2014), pp. 371–372

Z. Yuan, Y. Lu, Y. Xue, Droiddetector: android malware characterization and detection using deep learning. Tsinghua Sci. Technol. **21**(1), 114–123 (2016a)

Z. Yuan, Y. Lu, Y. Xue, Droiddetector: android malware characterization and detection using deep learning. Tsinghua Sci. Technol. **21**(1), 114–123 (2016b). https://doi.org/10.1109/TST.2016.7399288

M. Alshehri

J. Zhang, F. Zou, J. Zhu, Android malware detection based on deep learning, in *2018 IEEE 4th International Conference on Computer and Communications (ICCC), Chengdu, China*, (2018), pp. 2190–2194. https://doi.org/10.1109/CompComm.2018.8781037

Y.-l. Zhao, Q. Qian, Android malware identification through visual exploration of disassembly files. Int. J. Netw. Secur. **20**(6), 1061–1073 (2018)

H. Zhu, Y. Li, R. Li, J. Li, Z. You, H. Song, SEDMDroid: An enhanced stacking ensemble of deep learning framework for Android malware detection, in *IEEE Transactions on Network Science and Engineering*, (2020). https://doi.org/10.1109/TNSE.2020.2996379

Chapter 5
Exploring and Analysing Surface, Deep, Dark Web and Attacks

Jabeen Sultana and Abdul Khader Jilani

5.1 Introduction

The development of web-based media and versatile innovations changed web into all-inclusive wellspring of data. Whenever a search is made to get desired data using conventional search engines, data available at the surface level is only retrieved and the deeper hidden data is invisible. To overcome this and to satisfy the thirst for the desired data, deep web has to be accessed and to get data in more depth dark web needs to be accessed using specialized tools and technologies like TOR. The process to retrieve the deep data stored in the web and its applications are discussed and analysed here. Internet WWW is a data space that distinguishes web assets and reports through uniform asset finders interlinked by means of hypertext interfaces, which can be gained access through web. Web pages can be retrieved with the help of web browsers or using search engines like Google, Yahoo, etc. Internet is the basic connectivity to gain access to WWW and in turn WWW is a part of Internet. WWW relentlessly developed from its unique plan to fuse online media and client produced content. Web is the most noticeable part of the Internet. It contains web pages that incorporates text just as other multimedia data. Individuals are using web as a way to discover data about the subject of their advantage. They depend on web crawlers, for example, Google-the biggest search engine, Yahoo and Bing for accessing the desired data. They assess query items and utilize the most desired data retrieved by the web crawlers. On the off chance that either the ideal data isn't found or client isn't happy with the data given, at that point, the client can play out another hunt or refine their inquiry. The client ought not be over-burden with data. Looking

J. Sultana (✉) · A. K. Jilani
Department of Computer Science, College of Computer and Information Sciences, Majmaah University, Al Majmaah, Kingdom of Saudi Arabia
e-mail: j.sultana@mu.edu.sa; a.jilani@mu.edu.sa

© The Author(s), under exclusive license to Springer Nature Switzerland AG 2021
A. Bhardwaj, V. Sapra (eds.), *Security Incidents & Response Against Cyber Attacks*,
EAI/Springer Innovations in Communication and Computing,
https://doi.org/10.1007/978-3-030-69174-5_5

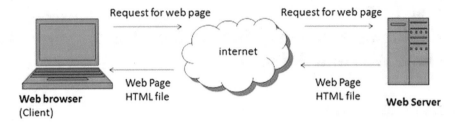

Fig. 5.1 General architecture of WWW

for wanted data in the web can be contrasted with the demonstration of hauling the net over the outside of a sea. Web crawlers are the core part of the Internet. The below Fig. 5.1 shows the architecture of WWW (Beshiri and Susuri 2019) (OBJ. Survey of current web architecture).

5.2 Search Engines Work

5.2.1 Web Browsers

Starts via looking through intensely utilized servers and the most famous web pages. Visits all the connections found in the web pages it peruses and sends them to the internet searcher's indexer through some process termed as web crawling, which proceeds naturally. To be found, a web page must be static and connected to different pages. All pages that are filed by a web crawler's browser programs are known as the noticeable web.

5.2.2 Invisible Web or Deep Web

Everything found by a web crawler's browser program cannot be shared by invisible web. Most website pages on the web are not filed via web crawlers. Some specialists gauge that as much as 75% of the whole web is undetectable web content and is stored in the invisible web. Web pages without any connections on them are called separated pages. Password ensured web pages. Web pages are produced from information warehouses. Dynamically web pages contain real-time content. Web pages that require an enrollment procedure to get access and web pages with non-html text, or any coding that a bug program cannot comprehend. The below Fig. 5.2 shows classification of web.

Internet is an immense archive that contains information in either surface web or deep web. Surface web is a segment of the web that can be gained access through connection crawling strategies. Crawling link is finding connected information

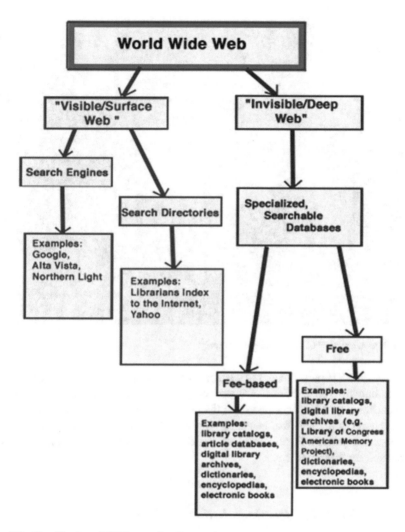

Fig. 5.2 Classification of WEB as surface/deep web

through hyperlinks from the main page of a particular domain space. Deep web is a part of the web that is avoided from the perspective of conventional web crawlers. It is a mother lode of unfathomable and fascinating data that is unreservedly open. Dark web is a little segment of the deep web whose content is covered up deliberately for some particular reason. Accessing dark web needs unique tools, for example, TOR, IIP and Freenet to gain access to the data stored there and requires the most significant level of encryption. The below Fig. 5.3 depicts the general view of surface, deep and dark web.

Data is considered as the fuel in today's world. Web has titanic measures of information. High dependency over search services helps to discover needed data

Fig. 5.3 General view of surface/deep/dark web

from web. Web is partitioned into surface web and deep web, as per the availability of the content. Web content that is promptly available via web crawlers is surface web and web content concealed from the perspective of web crawlers is deep web. It is segment of the web that is neither available by ordinary web indexes nor obvious to web browsers. For Instance, academic diaries, databases of users, web mail pages and service gateways.

5.2.3   What Is Hidden Deep Inside the Web?

The hidden things in deep web are profound web contents that include dynamic content produced on the fly. Unlinked content deals with web pages that are not connected to and by different pages followed by scripted content that are downloaded from web workers progressively. Content that restricts its entrance in fact is called limited access content and also content outside of http://. Secret key secured information generating from private web browsers and web content changing for various access settings including contents from blog that is composed however not yet distributed. The significance of searching deep web data underneath the surface web aids in getting access to gigantic data as 96% of data is hidden inside the deep web and it is freely available. Topic-explicit contents for state-of-the-art data are

Fig. 5.4 Clear view of surface/deep/dark web

available in these deep data warehouses. Great Quality data belonging — to different categories of data are available here in the deep web and the below Fig. 5.4 gives clear view of different kinds of web.

There are some difficulties in getting access to deep web data by conventional web crawlers because of huge size of data, non-filed, need fixed URLs. Independent, unconnected and sporadically disseminated data are continuously changing every now and then. Also, heterogeneous nature of data and to access particular data, need membership charges and thus how such data is hidden from web crawlers. Despite the fact that covered up, deep web data can be accessed straightforwardly by clients. Assuming web as a big sea, the surface web is the head of the sea that can be seen effectively or 'open'; the deep web is the more profound aspect of the sea underneath the surface; the dark web is the lower part of the sea, like a spot in deeper part of sea and this dark web data can be available by using extraordinary methods or tools like TOR (TOR Project n.d.).

5.2.4   Research on Deep Web and Dark Web

'Deep web' alludes to the basic information on the web that can neither be brought by conventional web crawlers nor obvious to web clients. The term 'deep web' was introduced by Michael K Bergman in 1990s due to its enormous size. It is quickly expanding at an exponential rate. Deep web is around 450–550 times bigger than the surface web in size and is both subjectively and quantitatively not quite the same as that of the surface web. Despite the fact that it is less gained access by the public, deep web is possibly significant, a secret stash of fascinating and amazing data with rich and fluctuated contents. Disclosing/introducing deep web offers enormous exploration beginnings in the region of data mining. The analysts have likewise incorporated the means and safety measures taken before the dark web was opened. Aside from that, the discoveries and the site joins/URL are likewise included alongside a portrayal of the locales (Rafiuddin et al. 2017). The impact of the dark web is more pertaining to various circles of society. It is given the quantity of every day mysterious clients of the dark web in utilizing TOR in Kosovo just as in the entire world for a while. The impact of shrouded administration sites is appeared and results are accumulated from Ahimia and Onion City Dark web's indexes. The namelessness isn't totally checked on the dark web. TOR devotes to it and has proposed to give unknown movements. The quantity of clients and in which place(s) they are located are described. The estimation depends on IP delivers as per nation codes from where comes the admittance to them and report numbers in total (Beshiri and Susuri 2019).

There is a need to investigate deep web with appropriate methods and techniques as deep web can be acquired by directing queries straightforwardly. Inquiring deep web web page 'each in turn' is difficult. It is unreasonable to revise same cycle over various deep websites. For all intents and purposes, it is difficult to revise whole measure on a sensible update to oblige dynamic deep web information. Besides, either sort of inquiries or data wanted by the client can't be normal. In this way, investigating deep web without suitable innovation is an incomprehensible errand. Customer side devices are sufficiently bad, and, just technically knowledgeable can get to deep web. Despite the fact that pre-collected storage facilities exist, they can't fulfil all information requests of the client. Despite the fact that a few advancements and methods are set up, they should be enhanced with a persevering investigation framework adaptable by the client to access deep web.

5.3   Trending Research

'Deep web and dark web exploration' is a procedure to investigate the information from the deep web dependent on the queries specified by the client. Booming deep web to surface is need of today on account of inborn characteristics related with deep web. Effective frameworks for investigating deep web and dark web are proposed and furthermore improved algorithms for ranking are recommended for effective pursuit. The dark web is essential for the web that isn't obvious to web indexes and requires the utilization of an anonymous program called Tor to access it. For instance, bitcoin exchange and silk trade are carried out through dark web (Bitcoin News n.d.).

5.4 Distinct Characteristics Between Deep Web and Dark Web

Dark web is a portion of the deep web that is deliberately covered up and open just through extraordinary programming, for example, TOR and I2P. The movements on the dark web are totally unknown and require most significant level of encryption. This portion of the deep web is called as 'Darknet' or 'Shadow Web'. Fundamental safety measures must be followed to get access to the contents of Dark Web.

Deep and dark web share some similar features in like manner: being unavailable by ordinary web indexes, absence of fixed URLs, spread in and around the Web. However, there is a significant difference. Deep web contents are not covered up purposefully and don't need extraordinary programming for their entry. They can be effectively open through ordinary channels whenever it is realized where to look separated from what to look. Deep web crawlers, invisible web catalogues, and Wikis can assist clients in getting access to deep web. The contents of dark web are covered up purposefully and the movements on dark web are totally unknown. Accordingly, deep web advancements have zero contribution with the dark web. In a nutshell, deep web however can't be gained legitimate access via web indexes, can be gained access straightforwardly by clients through querying interfaces.

Some criminal activities too take place using dark web. All dark web assets are not unlawful, shameless, or illicit. Writers utilize the dark web to impart more secure information passage with nonconformists. Non-profitable organizations utilize dark web to permit their labourers to associate with their home website, without informing that they are working with the association (BrightTALK n.d.). Surface web, deep web and dark web are interconnected ideas aside from gaining access to content availability and profundity of data. Surface web movements are profoundly made safe and secure about and those at deep web are covered up and concealed from mysterious tasks contents which are significant portion of dark web's movements. The surface web has lawful movements, while dark web houses dark and obscure movements. Data searchers should surf securely on the surface web, just swim in the deep web of approved websites for authentic reasons and dodge dark web for security reasons (Suneeta and Usha Rani 2017) (Fig. 5.5).

Conventional web crawlers can't test underneath the surface web and henceforth deep web seems, by all accounts, to be covered up. Deep web is a mother lode of data. Looking and crawling of the deep web is practically incomprehensible by existing URL-based web indexes, because of this out of reach nature of the deep

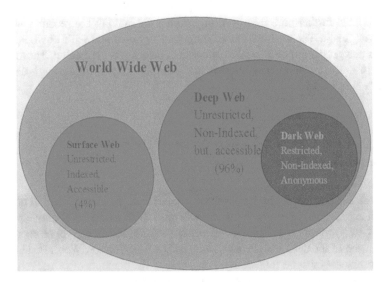

Fig. 5.5 Surface web, deep web and dark web interconnected

web. Despite the fact that extensive research is being done on deep web from the last denary, still there is a great carrier for incorporating creative works associated with deep web hashing, ranking and indexing. The above Fig. 5.5 depicts the interconnection of surface, deep and dark web.

5.5 Benefits of Deep Web

Specific focus is laid on desired contents centre—a lot of data analysed focusing on a specific subject or domain. Contains data that probably won't be accessible on the surface web. Allows a client to locate an exact response to a particular search. Allows a client to discover web pages from a particular date or time.

5.6 Deep Web Access Procedure in TOR

The Tor network is an encrypted network that allows anonymous access to the Internet for its users. The Tor network also hosts hidden services which constitute the infamous dark web. These hidden services are used to carry out activities that are otherwise illegal and unethical on the surface web. These activities include distribution of child pornography, access to illegal drugs and the sale of weapons. While Tor hidden services provide a platform for uncensored ventures and a free expression of thoughts, they are outnumbered by grey activities taking place.

Onion: Onion is a specific stream which hosts a suffix conveying mysterious concealed assistance reachable through the Tor organization. In view of innovations created by the US Naval Research Laboratory in 1990s to ensure US insight trade-offs on the web. Tor was first released to public in 2004. The Tor Project Inc. dispatched in 2006 as a non-benefit association. The motivation behind utilizing such a framework is to make both the data supplier and the individual getting access to the information harder to follow, regardless of whether by one another, by the middle organization, or by the outsiders. Onion addresses are 16-character non-memory helper hashes, bargained of alphabetic and numeric strings. The 'onion' name alludes to onion directing, the strategy utilized by Tor to accomplish a level of secrecy.

5.7   How TOR Works

The Tor secures you by bobbing your correspondences around a circulated organization of transfers run by volunteers all around the globe: it forestalls someone viewing your Internet association from realizing what destinations you visit, it forestalls the locales you visit from learning your physical area, and it lets you access destinations which are obstructed [TOR]. Pinnacle is an organization that upholds onion directing, an approach to help make your traffic unknown. Since the deep web is contained data that doesn't appear on web indexes, or has no space name vault, you should know precisely where you will arrive. The initial step is to download, instal Tor on your desktop, and run the Tor bundle. This will raise your new mysterious program by means of browser so as to begin.

5.7.1   TOR Offers Anonymity

Mysterious admittance to surface web administrations is possible through Tor. The Tor network gives an unknown access through the Tor organization. The Tor network 'leave hub' makes an interface with the surface web servers and provides anonymous entry to conceal the identity of both client and administrations. The Tor network gives total start to finish obscurity. Hides the character of both customer and administrator. The below Figs. 5.6 and 5.7 shows the working of Tor.

5.7.2   Measures to be Taken Before Gaining Access to TOR

On the off chance that you choose to investigate the deep web on your own volition, try to be cautious.

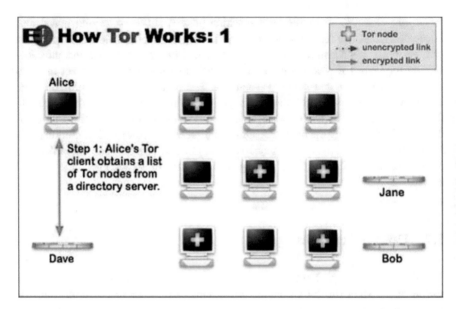

Fig. 5.6 How TOR Works-1

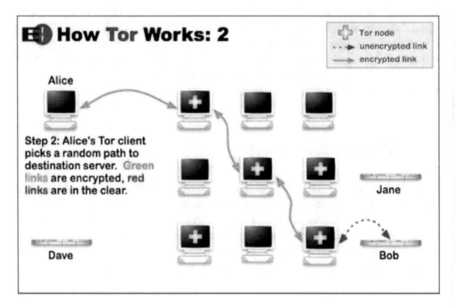

Fig. 5.7 How TOR Works-2

- Have hostile antivirus software to overcome the infections by the intruders.
- Be insightful about what joins you click. The deep web is an asylum for phishers.
- If you would prefer not to perceive any upsetting pictures or content, basically text contents as it were. The Tor network provides complete end-to-end anonymity and hides the identity of both client and server.

5.7.3   Attacks Inside the Dark Web

Checking the dark web secures against online protection dangers and gives a perspective in reality where cybercriminals accumulate to exploit your business. An ongoing phishing scam on the dark web found more than 160 million information records, taken from more than 12 organizations and were available to be purchased in open. These records contained individual information, including names, email locations and passwords. In another dark web find, more than 500,000 Zoom accounts were accessible to buy. The dark web is a hive of cybercrime activities taking place every now and then, a commercial centre used to sell individual information, cybercrime apparatuses and organization insight, including Intellectual Property (IP) and login certifications. This ordnance is then hence used to submit digital assaults. The below Fig. 5.8 gives overview of deep web markets.

The dark web doesn't simply contain taken information available to be purchased. It is a secret stash for the cybercriminal crew with all they require to run illegal activities. Concealed from ordinary web crawlers, the sites, discussions and commercial centres inside it contain the instruments and data expected to execute refined digital assaults. Recognizing what you are facing is a significant piece of battling cybercrime. This is the place observing and investigating comes in. The

Dark Web Markets

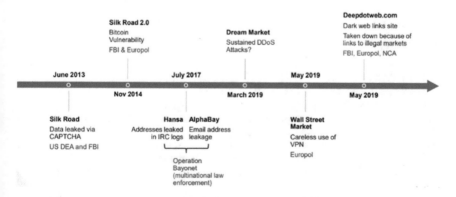

Fig. 5.8 Deep web markets

dark web, as different organizations, can be checked. The threat intelligence accumulated would then be able to be utilized to forestall the very wrongdoings that dark web propagates.

Since most of the people are Working from Home (WFH) we have to guarantee that we twofold down on our endeavours to forestall digital assaults. Observing the dark web gives your institute the danger knowledge expected to secure against digital assaults across even your all-encompassing home networks.

5.8   Conclusion

The dark web has many authentic web pages and is utilized by individuals, for example, columnists and law requirement. Notwithstanding, it has likewise gotten inseparable from cybercrime. The 225,000 or so sites, discussions, and so forth, that are inside the limits of the dark web, are just available utilizing master programs and web indexes, similar to the Tor. These tools give namelessness to clients who enter the dark web and go under the assurance of the 'Onion' network.

This secrecy is the thing that makes the dark web so attractive to cybercriminals. Inside the dark web, programmers and fraudsters offer various administrations and data, including: Tools to stock exchange: Cybercriminals can purchase and sell the apparatuses expected to do digital assaults. This incorporates leasing phishing packs to complete phishing efforts and Ransomware-as-a-Service to coerce cash. Chat rooms and discussions: Hackers and fraudsters trade knowledge on targets and update each other on strategies utilizing talk rooms. Hackers for enlist: Freelance gateways where cybercriminals offer their administrations. Data available to be purchased: The dark web has numerous commercial centres that sell copied information for deceitful use. Information can incorporate individual information just as organization information, for example, IP, source code, organization data and advanced endorsements.

References

A. Beshiri, A. Susuri, Dark web and its impact in online anonymity and privacy: A critical analysis and review. J. Comput. Commun. **7**, 30–43 (2019)

Bitcoin News. https://news.bitcoin.com/beginners-guide-buying-goods-darknet/

BrightTALK(n.d.),https://www.brighttalk.com/webcast/10813/376953/illuminating-the-dark-web

OBJS. http://www.objs.com/survey/WebArch.htm

M.F.B. Rafiuddin, H. Minhas, P.S. Dhubb, A dark web story in-depth research and study conducted on the dark web based on forensic computing and security in Malaysia, in *IEEE International Conference on Power, Control, Signals and Instrumentation Engineering, Chennai*, (2017), pp. 3049–3055

S. Suneeta, M. Usha Rani, Unveiling deep web, a high quality, quantitative information resource. IJLTET **9**(2), 167–174 (2017)

TOR Project. www.torproject.org

Chapter 6
Securing ERP Cyber Systems
by Preventing Holistic Industrial Intrusion

Sunil Kaushik

6.1 ERP Systems

The ERP systems provide the state of the operations to the whole organizations and are known for real-time visualization of operations and providing efficiency to the whole enterprise. ERP systems are tightly integrating the various business process and modules and collect, store, and manage the data of these processes. Various researchers have tried to provide the definition of Enterprise Resource Planning (ERP). Chung and Synder (1999) defined the ERP as a category referring to similar products under one bigger product and Klaus et al. (2000) defined it as a tool, which unifies the process and data to offer a business solution. Al-Mashari (2002) refers ERP system as the most innovative development in any sector. ERP an innovation as presented by Al-Mashari (2002) can be thought of as a software that integrates the business functions of various departments and organizations (Koch 2003). This definition is an extension of the definition given by Rooney and Bangert (2000) and is further supported by Hoch and Dulebohn (2012). Marnewick and Labuschagne (2005) and Aladwani (2001) referred ERP as business software package that integrates the majority of business processes of an organization and automates the system to make all departments use common data and practice. Hence, ERP not only serves as a platform to enable several departments efficiently and effectively but also provides an environment wherein all departments are connected to each other.

S. Kaushik (✉)
University of Petroleum and Energy Studies, Dehradun, India

© The Author(s), under exclusive license to Springer Nature Switzerland AG 2021   97
A. Bhardwaj, V. Sapra (eds.), *Security Incidents & Response Against Cyber Attacks*,
EAI/Springer Innovations in Communication and Computing,
https://doi.org/10.1007/978-3-030-69174-5_6

Ross et al. (2006) conveniently using the above definition defined ERP as a business management system that comprises of integrated set of comprehensive software modules. Rosa et al. (2012) alluded the ERP as COTS systems (commercial off-the-shelf) designed to integrate all core functions of an enterprise on a unified database, regardless of the type, size, or nature of the business and has further stated that ERP is an extension of the manufacturing resource planning (MRP). ERP systems are configurable system which not only integrates but flawlessly enables to share the information for various operational or management activities.

For this study, ERP can be defined as application conforming to following points (Fig. 6.1).

- Business management application
- Consistent and integrated data
- Cost effective and efficient
- Integrates the business process and brings best in class processes

Current Enterprise Resource Planning (ERP) system is the outcome of the continuous improvements done for five decades. The continuous improvement was due to management techniques and due to developments in the field of software and

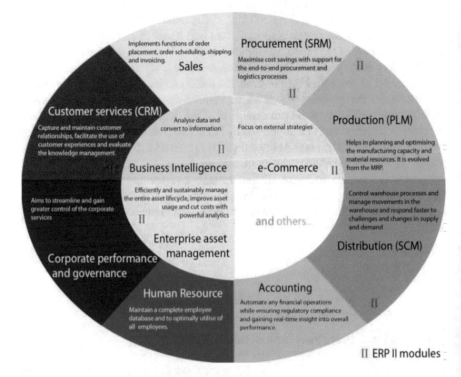

Fig. 6.1 Modules in ERP

hardware technologies. In the 1960s, corporates created their in-house software based on the classical/scientific inventory control techniques to automate their work and optimize the inventory (Bishnoi 2011). The 1960s saw the development of material requirements planning to calculate the material requirement according to master production schedule. MRP systems were further enhanced in the 1980s by including the processes to optimize the inventory by simulation and forecasting (Bishnoi 2011). The 1980s–1990s global competition, shortened product cycle, and customer focus led to the integration of various functional areas of an organization. The requirement for integration called for the requirements of Enterprise Resource Planning (ERP). The ERPs evolved contain all the back-office solutions such as financials, order management and distributions and human resources management system (Bishnoi 2011) and evolved to decision support systems. MRPs transformed into ERP to handle the competitive challenges thrown by industry (Davenport (1996). It can also be suggested that ERPs are evolved version of MRP/MRP II system (Al-Mashari 2002) to support various business vertical of an enterprise (Kennerley and Neely 2001).

ERP systems are created to handle integrated multiple functions (Fig. 6.1) in a complex corporate environment in an efficient way by providing the information-enabled environment (Amoako-Gyampah 2007). ERP implementation has helped the organizations to be customer centric and to reap the benefits of standardization, business process reengineering to take advantage of quality standards (Wortmann 1998). Kennerley and Neely (2001) have listed the following benefits of ERP implementation (Table 6.1).

Further, Turban et al. (2006) point out the reason of success of ERP and benefit is that ERP provides real-time picture of all the transactions happening across organization. ERP systems are able to providing significant tangible or intangible benefits by integrating various business functions. The intangible benefits are standardization and improved business processes, informed decisions, and better visibility of organization. ERP systems have flawlessly integrated the information from an entire enterprise (Bueno and Salmeron 2008). Gefen and Ragowsky (2005) add that ERP applications improve efficacy and value further by automation.

Table 6.1 Benefits of ERP Implementation

Increased coordination and control	The decision support systems have unburdened the administration departments and increased interaction between various parties has led to better control and coordination yielding high efficiency
Inventory optimization	Better control over stocks, assets, and consumables. An improvement in the procurement strategy renders better inventory optimization
Increased profitability	Informed decisions and better planning lead to increase in profitability
Better insight to suppliers	Suppliers get the view of bigger picture and this allows them to negotiate better for contracts
Capacity optimization	Larger view of orders and inventory levels enables organizations to plan the capacity well in advance

The main benefits of the ERP can be categorized into four parts:

1. Business benefits—Ability to view cross-functional data for informed decisions and actions. The streamlining of business processes can be seen as an added advantage.
2. Better integrations—Consistent data and data storage formats make it easier to integrate with third-party systems.
3. Technological benefits—One database and multiple technologies in one suite makes it easier to use and right technology is used for right work. Customizable and configurable processes make it easier to extend or scale the system.
4. Functional benefits—Preloaded process makes it easier to chalk out the best process for any organizations. The processes built by taking best practices of mature organizations can benefit even a new organization.

ERP systems are slowly evolving into cyber physical systems to cater to the need of Industry 4.0. Cyber physical systems are the systems that integrate the software, network, and physical systems. In nutshell, it can be said that industry 4.0 is looking to integrate the payment systems, controlling systems, supply chain systems, or manufacturing systems. Industry 4.0 is delegating the backward integration related tasks and forward integration tasks to vendors and customers, respectively. This has carved out the need of the self-service modules in ERP wherein vendor and customers can login to ERP over internet and take actions. This exposure to internet, increasing number of transactions and systems into ERP or CPS are exposing the ERP to cyberattacks.

6.2 Integrations in ERP

ERP implementation projects are large-scale software projects, and are often incomplete without the box functionalities called customization (Luo and Strong 2004). The product is usable only after customizations (Aydin 2012; Aversano and Tortorella 2013) have been built and integrated. Customizations could be business process customizations which have low impact on the end product (Parthasarathy and Daneva 2014) or system customizations fulfilling the business requirements which have high impact on quality and cost because complete software development processes (requirements analysis, software design, coding, and testing) are carried out. ERP systems are complex commercial off-the-shelf (COTS) software packages and their fitment into organization's requirement is always a challenge (Vilpola et al. 2006). The changes to fit into organization's working style are quite underestimated (Van Stijn and Wensley 2001). Organizations tend to fill the gap between ERP and their processes by either altering their business processes or customizing the ERP system by rewriting part of the delivered software, or interfacing to an external system (Fryling 2010). The various types of customizations are given in Table 6.2.

Table 6.2 Typology of ERP tailoring types

Tailoring type	Description	Layers involved
Configuration	Parameter setting (or tables), to choose between different processes and functions in the software package to execute	All layers
Bolt-on	Any third-party software implemented in tandem of ERP to provide industry-specific solution	All layers
Screen masks	Representing the data in various screen formats for various users	Communication layers
Extended reporting	Various reports showing same data in different formats	Application layer and/ or database layer
Workflow programming	Creating non-standard workflows	Application layer and/ or database layer
User exits	Programming of additional software code in an open interface	Application layer and/ or database layer
ERP programming	New functionalities on top of source code in the vendor's own language	All layers
Interface development	Programming of interface to legacy system or third-party products	Application layer and/ or database layer
Package code modification	Changing the source codes ranging from small changes to change whole modules	Can involve all layers

6.3 Challenges in ERP Systems

ERP systems as other system have benefits and associated challenges. These challenges could be running the ERP at optimum performance and provide correct data for informed decisions to secure the data from unauthorized access. Few of the challenges are listed below:

1. Application Architectures: ERP contains various modules and these modules are separate applications in themselves. ERPs have come into existence over years and few of the modules are homegrown or are acquired through acquisitions. However, ERP providers try to have a consistent architecture across the modules but few modules are always exception and have different architecture. Hence, these different applications pose a challenger in security (Horowitz-Kraus 2015).
2. Customizations: Any ERP implementation is incomplete without customizations. These customizations are provided by the system integrators and are not provided by the ERP provider. These customizations, however, try to follow the API given by ERP provider but have the own code written. However, industry tries to follow a standard guideline but these customizations pose a risk to security of the customizations (Michelberger and Horváth 2017).
3. Third-party integrations: ERP systems have to work with various third-party software and machinery to automate. The code of these software or machinery may not be that secure as that of the ERP. Data of the ERP is exposed to the

third-party software for viewing and updating. The source code of the third-party application is usually not shared with the enterprise and is mostly on cloud. This further aggravates the situation and exposes the system to chain reaction (Pavel and Evelyn 2017).

4. Security of data: ERP frameworks are such an extensive archive of various types of information that they can hugely affect security issues. Data security is a concern for inside and outside users. Data should be visible and updated by the authorized user else, it can create a huge distress inside the enterprise and can share the vulnerabilities outside enterprise (Geum et al. 2017).

5. Outside and internal data privacy issues: The issues for privacy may occur and be more crucial and tough to control within the company. The segregation of duties should be performed to restrict the rights to view or update not only the data and configurations. Sometimes the customization code does not take care of the segregation of the duties and sometimes the custom code has errors that may hit the accounting badly and may expose the whole system for any cyber intrusion (Hustad et al. 2016).

6. Change Management: The implementation of the ERP takes care of design and functional requirements and nonfunctional requirements such as performance. The security aspect is never taken into consideration during implementation or design phase. The security aspect of the ERP is never coded and tested for vulnerabilities. Hence, all these aspects should be considered well by the reviewer (Orougi 2015).

7. Different user id for ERP and Systems: Most of the enterprise use ERP as a separate system and do not integrate it with the user id of the domain. These multiple user ids expose the risk of any user id being exploited by internal as well as external users. The separate user id of the system may not have the password policy same as that of the LAN or application domain. The password without policies are easy to guess and never expire. The Single Sign On or Identity Access Management tool should be used to take care of these or a complex password policy should be set for the ERP systems (Chavan and Nighot 2016).

6.4 Cyber Threats on ERP Systems

ERP systems with the advancement of technologies are reaching to new level of functionalities and are becoming susceptible to the various cyberattacks such as phishing and DDoS. These attacks indulge in destroying the physical infrastructure, getting access to sensitive data, and doing fraudulent transactions (Venkadasalam 2015). These attacks and compromise on data result not only in financial loss but also in loss of trust from vendor and customer. These results are catastrophic in nature (Mangan and Lalwani 2016).

The impact of any kind of cyber intrusion has increased by 200%, which is expected to be 300% in the next five years. Cybersecurity breach poses a dynamic challenge to businesses and threatens their smooth operations and competitive advantage. Despite widespread attention to the dangers of cyberattacks, many companies are not well equipped to address the issue (Calatayud et al. 2019). More often small businesses do not have the resources in place to protect themselves (Singh et al. 2019). Some businesses are more susceptible to intrusions than others, but usually most are exposed to potential attacks. Enterprises need to be planned in cyber defense and build a robust system that minimizes the impact of cyberattacks (Shivajee et al. 2019). Enterprises and governments are taking many procedures in order to thwart these cybercrimes. Despite all these, cybersecurity is a point of concern for all. The impact of cybersecurity worsens if the organization has ERP underneath. ERP systems are more vulnerable to cyberattacks and the impact is disastrous since it contains the vital information such as bank account of organization, suppliers, and customers, or the other trade secret information which might prove fatal if it is disclosed in public. In addition, ERP contains that data of the organization financial health which if compromised can affect the organization's performance in share market (Woods and Bochman 2018).

ERP can be seen as the technology that drives the economy and affects enterprise agility. ERP system, in future, will become a system that integrates heterogeneous, intelligent and provide a level of knowledge. Researchers have been discussing the ERP security issues for some time and few solutions have been provided with the assumptions that ERP system are closed environment (Mangla et al. 2016). The future ERP system may have the following features:

- Heterogeneous: This means that the components or functions from different technology coexist and cooperate in a system. It has two prerequisites: modularity and integration. This requires stronger communication platforms to support heterogeneous functions and applications on different unrelated technology (Saunders 2014).
- Collaborative: ERP systems collaborate the enterprise intrinsic processes and the enterprise extrinsic processes. The processes such as accounting of payables, receivables, and assets can be referred as intrinsic process while process ranging managing warehouse and inventory to delivery of the finished goods can be considered as the extrinsic process. Users using intranet work upon most of the intrinsic processes. Extrinsic processes are accessed beyond four walls of the enterprise where intranet is often unavailable so they are done through the internet. ERP seamlessly combines these two kind of processes without leaving any trace. The operational efficiency of an enterprise lies in the extrinsic processes (Terminanto 2014).
- Intelligent and Knowledge based: In future, ERP system is expected to do much more than recording transactions and reporting. ERP will be needed to do the predictive analytics, what-if analysis to provide a recommendation for tactical or strategic transformations.
- Mobility: Future ERPs will be able to provide all the features and functionalities on the go to the mobile workforce (Zare and Ravasan 2014).

The current methods, logging of event and access control, are going to be insufficient and ERP providing companies will fail in securing the data of company stored in ERP system. Putting extra controls on access system will not only make ERP system slow but will also prove to be strenuous in managing the enterprise with dynamic user base. In addition, logging all the transactions or enabling the trace for each user will also make the system slower. Hence, the current methods of securing the ERP are going to be the roadblock in enterprises operational efficiency and agility. This requires looking for a security system for ERP, which aligns with enterprise's goal generally, improved security solutions call for elevated cost and may affect the performance of a system. In this paradox, enterprises usually go for solution that does not hit the performance of the current systems and is available at lower cost. This is in line with the philosophy, which made them implement ERP (Petrasch and Hentschke 2016).

From Panaya's internal survey of over 200 top ERP customers, 70% of the responding enterprises skip security and compliance audits of their ERP system. Despite this, 60% of IT security professionals fear that the impact of an attack on their business applications would be catastrophic. The global cost of cybercrimes last year was reported at $600 Billion according to their recent report. In addition, while the costs are rising, so are the number of attacks. The current situation of the ERP security with growing number of incidents to steal the enterprise data requires ERP security to be revisited. This situation becomes more critical when an enterprise plans to go for Internet of the things (IOT) based solutions.

6.5 Potential Solutions

With the exposure of ERP on the internet for the external uses, the threat to the system has increased manifold. Few network topologies and architecture are discussed in the subsequent subsections. All of these architectures require the ERP in a de-militarized zone (DMZ) and firewalls at different levels to ensure that the only legally allowed traffic enters, crosses the firewalls and interacts with ERP. This architecture also would need an external application tier for restricting the access to a limited set of users logging in via the Internet.

6.5.1 Terminology

Below are definitions of some of the terms that are used in this document:

- Firewall: Network firewalls control access between the internet and a corporation's internal network or intranet. Firewalls define which internet communications will be permitted into the corporate network, and which will be blocked. A well-designed firewall can foil many common internet-based security attacks.

- DMZ: It stands for De-Militarized Zone, which consists of the segments of a corporate network between the enterprise intranet and the Internet. The main advantage of a well-configured DMZ is improved defense. In case of a security incident, only the area enclosed within the DMZ is revealed to possible damage, while the internal intranet of enterprise remains untouched and protected.
- Load Balancer: Spread an application's load over many identical servers. This distribution ensures steady and stable availability of application even when any of the server fails.
- Reverse Proxy: A reverse proxy server is a middle server that rests between a client—requestor—and the web server. The reverse proxy server in turn requests the web server on behalf of requestor client.
- Service: Functional set of Oracle E-Business Suite application processes running on one or more nodes.
- Node: Server that hosts the ERP application runs the processes or database processes. In a single node installation of ERP, all the processes including the database processes run on one node whereas in a multi-node installation, based on the configuration, various processes run on multiple or a designated node.
- Internal Applications Tier: It is referred as the node configured to enable internal users to access ERP. It runs the following major application services:

  - Web and Forms Services
  - Workflow Administration service
  - Node Manager
  - HTTP Server
  - Concurrent Manager Services Reports
  - Discoverer Services

- External Applications Tier: Server configured for external users for accessing ERP. It runs the following HTTP Server Web server components like node manager, managed servers, etc.
- Primary Application Tier Node (or master node): Application technology stack for Oracle E-Business Suite 12.2 utilizes WebLogic Server. The primary application tier node is the node that is running the WebLogic administration server.
- Secondary Application Tier Node (or slave Node): Application technology stack for Oracle E-Business Suite 12.2 utilizes WebLogic Server. A secondary application tier node or slave node is a node in a multi-node deployment of Oracle E-Business Suite. The slave node does not run the WebLogic administration server.
- Shared Application Tier File system: An application tier file system consists of APPL_TOP file system (APPL_TOP and COMMON_TOP directories).
- Application tier technology stack file: Instance Home (INST_TOP) file system. Each application tier and each file system edition (run and patch) has a unique Instance Home associated with it. In a shared application tier file system, the APPL_TOP, COMMON_TOP, Web Server, and Web Tier Home file systems are

mounted on secondary application tier nodes either from the primary application tier node or from an NFS server.

- Multiple Domains: Allow different users the ability to access ERP via different web entry points.
- URL Firewall: White list of URLs, for the externally exposed ERP Modules, that may be accessed from the Internet.

6.5.2 An Internal and External Application Tier with DMZ

This is the simplest configuration for any enterprise system to be accessible over the internet. Diagram shows two tiers:

1. Internal Application tier: This is configured on intranet to be used by the internal application users.
2. External Application Tier: Here external application tier configured on DMZ enabling users to access the system on the internet.

The external application tier configured in DMZ. The traffic from internet is marked as HTTP/HTTPS and the database commute is shown as SQLNet in the below diagram. In addition, the web servers are marked as WLS and NM is referred here as an option utility that runs as separate process from web server and allows performing common operations tasks for a managed server. The architecture diagram shows that any traffic to the ERP has to first go through the DMZ firewall, which allows only the trusted URLs to access the external application tier, which contains required forms and access to web servers. The access of the form is regulated by the ERP's access-based securities. The data retrieval and update is done using the trusted database connection which connects to database within Intranet. However, in case of traffic from intranet is routed directly to internal application tier, which has form, reports, and concurrent manager. The given configuration is very simple however; it does not allow sharing of file system and hence adds on the cost of the implementation (Fig. 6.2).

6.5.3 An External Application Tier with Reverse Proxy in DMZ

Configuration defined in the previous section can be augmented with the reverse proxy server in the DMZ after the external firewall. The external application tier of the application is put behind one more firewall usually called internal firewall and

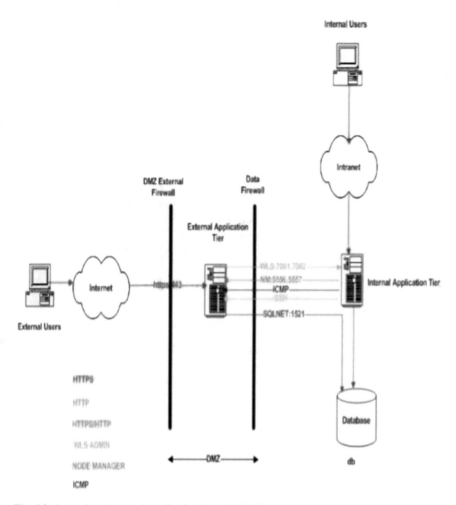

Fig. 6.2 Internal and external application tier with DMZ

talks to the internal application tier. The external application tier restricts the access over few functions for user logging outside the enterprise intranet/internet and the details of the external application tier are masked using the reverse proxy server. The reverse proxy server not only restricts the access to only selected set of internet URLs but also it terminates the SSL connection. This configuration like previous configuration also does not allow sharing of file system and hence adds on the cost of the implementation (Fig. 6.3).

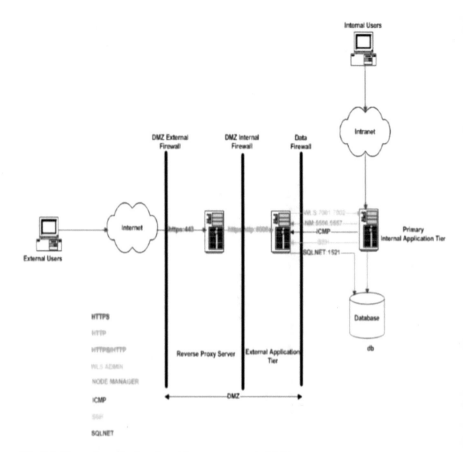

Fig. 6.3 External application tier with reverse proxy in DMZ

6.5.4 DMZ Sharing the File System with Application Tier on Internet

This architecture is an extension of the external application tier on DMZ architecture with a load balancer. The load balancer before the external application tiers. The load balancer is used to regulate and redirect the load from the users from the internet. These users could be external or internal and load balancer redirects them based on the type to different application nodes. This architecture benefits in terms that file systems are shared across all nodes which reduces the further load on the system to migrate the sanitized from external files to internal files which can be used by both external and internal users. In addition, it not only stops the requirements SSH connectivity to cross firewall but also it stops the requirement of opening a port of the web server over the firewall (Fig. 6.4).

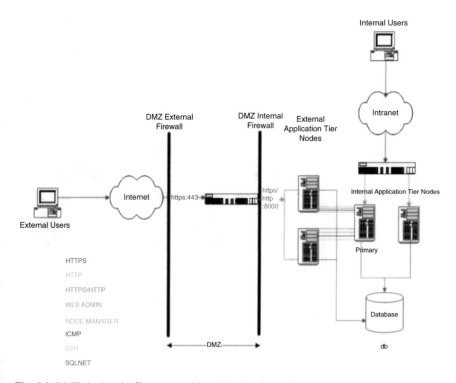

Fig. 6.4 DMZ sharing the file system with application tier on internet

6.5.5 Hybrid Setup

This architecture includes multiple external application tier nodes and multiple internal application tier nodes. These external and internal application tier nodes share common file system across nodes. The external application tier is set up in the DMZ, which uses a separate file system. In addition, to improve the load of various users a load balance is also set up in front to appropriately assign the traffic from internet to different nodes based on the access level. This configuration requires a separate level of firewall before the load balancer. The SSH connectivity and web server port needs to be opened over the firewall because of load balancer situated in firewall. In addition, a reverse proxy server can also be set up in front of load balancer within the firewall to provide more security to the system and to terminate the SSL connection (Fig. 6.5).

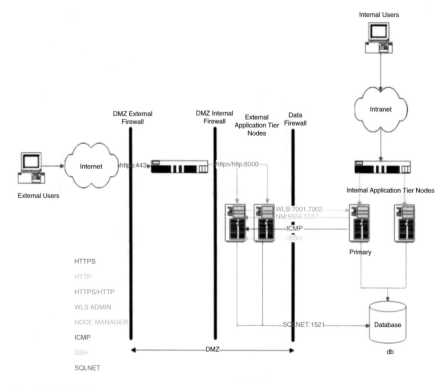

Fig. 6.5 Hybrid Setup

6.5.6 Few Precautions

1. **Set up the trust level:** If any of the self-service or over the internet module is implemented, then the ERP provider given trust level should be set up. These can be derived from the implementation guides provided by the ERP Provider.
2. **Update Security Patches:** All ERP providers keep on testing the product against the latest vulnerabilities and keep on releasing the patches. These patches should be applied to update in the ERP instance.
3. **Cached Object:** Web servers should be configured with the "Time to Live" settings. These settings enable to invalidate the cached objects on required intervals.
4. **Setting up HTTP and HTTPS Configuration:** Internet is always inflicted with performance issues because of unknown reasons. Hence, to increase the performance, public pages can be deployed in HTTP and sensitive data should be set up on HTTPS only.
5. **Forward Proxy Servers:** Forward proxy servers should be configured to handle outbound DMZ-to-Internet and outbound DMZ-to-Intranet HTTP traffic.

6.6 Conclusion

Organizations are storing various financial and operation data into ERP system. Hence security of the ERP systems has become major concern for organizations. Traditionally, ERP systems have been looking at the security from responsibilities, role-based access and segregation of duties perspective. In addition, with inclusion of self-service or internet module to access ERP from outside of enterprise has made it prone to cyberattacks like any other application. Hence, a framework or techniques are needed to stop the malicious attacks on the ERP. This chapter discussed the few architecture to secure ERP using the current infrastructure of any enterprise and provide a secure system at no extra cost.

References

A.M. Aladwani, Change management strategies for successful ERP implementation. Bus. Process. Manag. J. **7**(3), 266–275 (2001)

M. Al-Mashari, Enterprise resource planning (ERP) systems: A research agenda. Ind. Manag. Data Syst. **102**(3), 165–170 (2002)

K. Amoako-Gyampah, Perceived usefulness, user involvement and behavioral intention: An empirical study of ERP implementation. Comput. Hum. Behav. **23**(3), 1232–1248 (2007)

L. Aversano, M. Tortorella, Quality evaluation of floss projects: Application to ERP systems. Inf. Softw. Technol. **55**(7), 1260–1276 (2013)

A.O. Aydin, A new way to determine external quality of ERP software, in *Computational Science and Its Applications—ICCSA 2012*, (2012)

V. Bishnoi, Critical success factors for implementing ERP software in multinational companies operating in India. Yojna **2**(1), 9–12 (2011)

S. Bueno, J. Salmeron, TAM-based success modelling in ERP. Interact. Comput. **20**, 515–523 (2008)

A. Calatayud, J. Mangan, M. Christopher, The self-thinking supply chain. Supply Chain Manag. **24**(1), 22–38 (2019)

A. Chavan, M. Nighot, Secure and cost-effective application layer protocol with authentication interoperability for IOT. Proc. Comput. Sci. **78**, 646–651 (2016). https://doi.org/10.1016/j.procs.2016.02.112

S. Chung, C. Synder, ERP initiation—A historical perspective, in *Proceedings of America Conference on Information Systems*, (1999)

T.D. Davenport, Holistic Management of Mega Package Change: The Case of SAP, in *Association for Information Systems Annual Conference, Phoenix, Arizona*, (1996)

M. Fryling, Estimating the impact of enterprise resource planning project management decisions on postimplementation maintenance costs: A case study using simulation modelling. Enterpr. Inform. Syst. **4**(4), 391–421 (2010)

D. Gefen, A. Ragowsky, A multi-level approach to measuring the benefits of an ERP system in manufacturing firm. Inform. Syst. Manag. **22**, 18–25 (2005)

Y. Geum, M. Kim, S. Lee, Service technology: Definition and characteristics based on a patent database. Serv. Sci. **9**(2), 147–166 (2017). https://doi.org/10.1287/serv.2016.0170

J.E. Hoch, J.H. Dulebohn, Shared leadership in enterprise resource planning and human resource management system implementation. Hum. Resour. Manag. Rev. **23**, 114–125 (2012)

T. Horowitz-Kraus, Improvement in non-linguistic executive functions following reading acceleration training in children with reading difficulties: An ERP study. Trends Neurosci. Educ. **4**(3), 77–86 (2015). https://doi.org/10.1016/j.tine.2015.06.002

E. Hustad, M. Haddara, B. Kalvenes, ERP and organizational misfits: An ERP customization journey. Proc. Comput. Sci. **100**, 429–439 (2016). https://doi.org/10.1016/j.procs.2016.09.17

M. Kennerley, A. Neely, Enterprise resource planning: Analyzing the impact. Integr. Manuf. Syst. **12**(2), 103–113 (2001)

H. Klaus, M. Rosemann, G. Gable, What is ERP? Inform. Syst. Front. **2**(2), 141–162 (2000)

C. Koch, *The ABC of ERP* (Enterprise Resource Planning Research Centre. CIO, 2003)

W. Luo, D. Strong, A framework for evaluating ERP implementation choices. IEEE Trans. Eng. Manag. **51**(3), 322–333 (2004)

J. Mangan, C.L. Lalwani, *Global Logistics and Supply Chain Management* (Wiley, Hoboken, NJ, 2016)

S.K. Mangla, K. Govindan, S. Luthra, Critical success factors for reverse logistics in Indian industries: A structural model. J. Clean. Prod. **129**, 608–621 (2016)

C. Marnewick, L. Labuschagne, A conceptual model for Enterprise Resource Planning (ERP). Inf. Manag. Comput. Secur. **13**(2), 144–155 (2005)

P. Michelberger, Z. Horváth, Security aspects of process resource planning. Polish. J. Manag. Stud. **16**(1), 142–153 (2017). https://doi.org/10.17512/pjms.2017.16.1.12

S. Orougi, Recent advances in enterprise resource planning. Account. Forum, 37–42 (2015). https://doi.org/10.5267/j.ac.2015.11.004

S. Parthasarathy, M. Daneva, Customer requirements based ERP customization using AHP technique. Bus. Process Manag. J. **20**(5), 730–751 (2014)

J. Pavel, T. Evelyn, An illustrative case study of the integration of Enterprise resource planning system. J. Enterpr. Resour. Plann. Stud., 1–9 (2017). https://doi.org/10.5171/2017.176215

R. Petrasch, R. Hentschke, Process modeling for industry 4.0 applications: Towards an industry 4.0 process modeling language and method, in *Computer Science and Software Engineering (JCSSE), 2016 13th International Joint Conference*, (2016), pp. 1–5

C. Rooney, C. Bangert, Is an ERP system right for you? Adhes. Age **43**(9), 30–33 (2000)

W. Rosa, T. Packard, A. Krupanand, J.W. Bilbro, M.M. Hodal, COTS integration and estimation for ERP. J. Syst. Softw. **86**, 538–550 (2012)

J.W. Ross, P. Weill, D. Robertson, *Enterprise architecture as strategy* (Harvard Business School Press, Boston, MA, 2006)

L. Saunders, Linking resource decisions to planning, in *New Directions for Community Colleges*, vol. 2014, (2014), pp. 65–75. https://doi.org/10.1002/cc.2012

V. Shivajee, R.K. Singh, S. Rastogi, Manufacturingconversioncostreductionusingquality control tools and digitization of real-time data. J. Clean. Prod. **237**, 117678 (2019)

R.K. Singh, P. Kumar, M. Chand, Evaluation of supply-chain coordination index in context to industry 4.0 environment. Benchmarking (2019). https://doi.org/10.1108/BIJ-07-2018-0204

A. Terminanto, Forecast to plan cycle in Oracle E business suite (case study automotive company). Adv. Sci. Lett. **20**(1), 203–208 (2014). https://doi.org/10.1166/asl.2014.5279

E. Turban, D. Leidner, E. Mclean, J. Wetherbe, *Information technology management: Transforming organisations in the digital economy* (Wiley, New York, 2006)

E. Van Stijn, A. Wensley, Organizational memory and the completeness of process modeling in ERP systems: Some concerns, methods and directions for future research. Bus. Process Manag. J. **7**(3), 181–194 (2001)

S. Venkadasalam, Linear programming: An alternative Enterprise Resource Planning (ERP) in higher learning institution. J. Bus. Econ. **6**(9), 1633–1637 (2015). https://doi.org/10.15341/jbe(2155-7950)/09.06.2015/010

I. Vilpola, K. Mattila, T. Salmimaa, Applying Contextual Design to ERP System Implementation, in *CHI, Experience Report, Montreal, Quebec, Canada*, (ACM, New York, 2006), p. 147152

B. Woods, A. Bochman, *Supply Chain in the Software Era* (Atlantic Council, Washington, DC, 2018)

J.C. Wortmann, *Evolution of ERP systems. Strategic management of the manufacturing value chain* (Kluwer Academic, Boston, MA, 1998), pp. 11–23

A. Zare, A. Ravasan, An extended framework for ERP post-implementation success assessment. Inf. Resour. Manag. J. **27**(4), 45–65 (2014). https://doi.org/10.4018/irmj.2014100103

Chapter 7
Infrastructure Design to Secure Cloud Environments Against DDoS-Based Attacks

Akashdeep Bhardwaj, Sam Goundar, and Luxmi Sapra

7.1 Introduction

Internet has become the key driver for an organization's growth, brand awareness, and operational efficiency. Unfortunately, cyber-terrorists and organized criminals recognize this fact as well. Cybercriminals perform denial of service attacks in order to deny legitimate authentic users access to their authorized hosted services, by causing web sites to perform slowly and deny access to corporate network and data. Ensuring safety and security of information and communication technology and infrastructure has become a persistent race between the cyberattackers or black hats and the ethical hackers or defenders. To prevent such cyberattacks on cloud platforms, web portal hosting, service providers, and Internet data carriers ensure the highest focus and priority to these challenges. Since new vectors of attack and emerging risks are on the increase, businesses must defend IT infrastructure from sophisticated attack methods. Today, cyber-threats take on a number of styles and proportions. Due to the growing usability of botnet, big attacks have been recorded more regularly, with even 20 Gbps. Besides an uptick in pace, attacks have also been streamlined and durable. The essential resources are impacted such as network connectivity, session functionality, application support capability, or the back end database response. Layer 7, for example, application attacks are often more target-oriented and mostly consist of what seems genuine traffic making it impossible to spot.

A. Bhardwaj (✉)
School of Computer Science, University of Petroleum and Energy Studies, Dehradun, India

S. Goundar
Centrum Business School, Lima, Peru

L. Sapra
Chitkara University, Chandigarh, India

© The Author(s), under exclusive license to Springer Nature Switzerland AG 2021   113
A. Bhardwaj, V. Sapra (eds.), *Security Incidents & Response Against Cyber Attacks*,
EAI/Springer Innovations in Communication and Computing,
https://doi.org/10.1007/978-3-030-69174-5_7

Since cloud computing utilizes the Internet as the primary communication medium, traffic from unsecure Internet locations flow into public data centers, on-premise corporate systems, and home computers. This exposes data and computing resources to external threats and cyberattackers. Security vulnerabilities and problems need to be discussed and handled in order to harness the full value of cloud infrastructure to prosper. Cloud vendors and organizations incorporate hacker detection and firewall services in data centers, where the antivirus and personal firewall are mounted on home users on desktops and laptops to fight this attack.

However, when receiving and sending traffic to the unsecure Internet, inbound access to the corporate web portals, hosted applications, email servers, and computing systems is essential. Cloud service providers and corporate organizations open access to ports on the edge and network devices. In doing so, the corporate resources, devices, and data are again exposed to the cyber-threats at network and application level. Data centers, systems, and infrastructure devices are constantly targeted by cyberattackers and remain under constant threats primarily from malware and distributed denial of service attacks. Even though these attacks are not unique threats in itself, the detection and mitigation continues to be the top security challenge. These attacks constantly evolve into a new threat levels impacting the security solutions implemented in cloud and on-premise hosting environments. Each domain and segment is vulnerable to malware and cybersecurity DoS attacks which continue to increase in size, sophistication, and frequency. Thus there is an immediate need to have a secure security architecture that can detect and mitigate the network and application layer attacks and normalize the web traffic and ensure the hosted applications are available at all times.

In recent times, distributed denial of service and malware attacks have become the main security threats as described by Choi et al. (2010). DDoS attacks are executed to disable networked circuits, server systems by limiting access to them and termed as DoS attacks. These deny users the access to applications like Email, Chat, E-commerce or Banking, or hosted cloud services like SaaS, PaaS, or IaaS cloud services and computing resources like network or VoIP infrastructure. The attacks are performed from a single source address as described by Deshmukh and Devadkar (2015) and illustrated in Fig. 7.1 below.

7.1.1 Distributed Denial of Service Attacks

DDoS-based attacks have begun on Gaming and Casino platforms, recent cyber-assaults are also being used as a diversionary weapon to manipulate intellectual property and data for political purposes, financial profits, and otherwise. This cyber-attacks then escalate the denial of service attack by initiating a flooding attack from several thousand nodes, bombarding the target with malformed requests for information and data packets, thus crippling the networks and disabling normal operations. The attacks are referred by Mishra et al. (2011) and Anwar et al. (2015) as DDoS attacks. Figure 7.2 below illustrates the use of botnets sending amplified

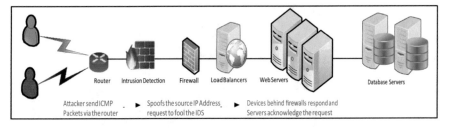

Fig. 7.1 Denial of service attack process

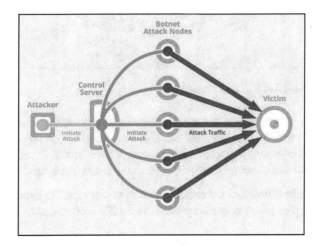

Fig. 7.2 DDoS attack using botnet nodes

requests and turning the DoS attack into a DDoS flood attack. The attacker exploits vulnerable systems across geographies by compromising them with a malicious payload. This payload infects the end user systems with a malware application which enables the attacker to gain remote access with command and control capabilities.

This is performed without the knowledge of the users with the intent to have the target services, hosted web applications unavailable to the authorized users and for cloud computing security issues. This is presented by Wong and Tan (2014), Zargar et al. (2013) with such cyberattacks being performed by sending a flood of network packets, data, or transaction requests over the Internet. These are sent from multiple locations and multiple systems at the same time. The infected and compromised user systems or nodes are referred to as Zombies or Bots which further compromise other user systems. The flood of compromised systems working as a group is known as botnets and controlled by a single attacker performing the attack sequence as shown in Fig. 7.3. DDoS attacks present a high priority risk for cloud service providers and cloud service consumers with regard to the hosted infrastructure for managing the service level agreements, cloud service delivery, cloud availability and avoiding any collateral damages.

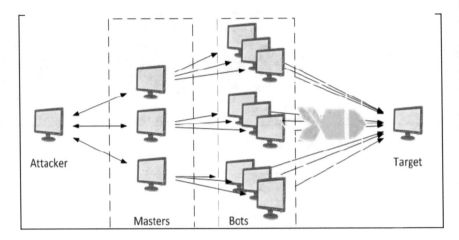

Fig. 7.3 Distributed denial of service attack

Apart from the cloud infrastructure and services unreachable to actual consumers resulting in rental losses, increased delivery cost, and harm to reputation, which further lead to legal and financial consequences, DDoS attack directly impacts the cloud providers and cloud service consumer in the below mentioned ways.

- Resource exhaustion like overwhelming and consuming the bandwidth capacities of Internet pipes or making the server CPU and memory to exceed the capacity.
- Triggering fallbacks to have the intrusion detection systems or web application firewalls alter from their filer-mode to log-only-mode.
- Exploitation of end user accounts with lockouts by repeatedly attempting logon access with invalid credentials.
- Cloud portal access disruption by crashing the web application process by attacking vulnerability is in the application code or altering user types to an invalid type and making it incorrect to input data for the legitimate user.
- Camouflage the real attack motive by diverting the security team attention acting as a smoke screen to steal data or hijack other services.
- Pushing malware which affects the user access and data by opening up sockets and triggering errors in the micro codes.

7.1.2 Types of DDoS Attacks

DDoS attacks are broadly categorized into three main types of attacks depending on the area of cloud infrastructure on which the cyberattack is focused. These attacks are described in the below section as volumetric, application layer and reflector DDoS attacks.

7.1.2.1 Volumetric DDoS Attacks

These attacks are network bandwidth attacks attempting to overpower and saturate the network bandwidth capacity (Giga bites per second). These originate from a botnet at most times over the Internet. Internet consists of a vast number of individual networks, interconnected to each other with the large, well-connected networks providing access to smaller networks. The connections between these networks have a finite amount of bandwidth capacity. This capacity is often fixed due to technical or contractual limitations. Regardless of the limitation, most links cannot be trivially upgraded to a higher capacity without incurring substantial cost in terms of both time and money. These attacks are performed on layer 3 and 4 protocol layers by flooding and consuming the network bandwidth to the point where access to the hosted resources is rendered inaccessible. TCP/UDP/ICMP floods and spoofed packet attacks are typical examples of such attacks and these are referenced as mentioned below from the Incapsula DDoS Attack glossary (2017).

- TCP Flooding starts by spoofing IP address in SYN packet data packet's header is sent to the header of the data packet is sent to the security monitoring systems.
- Attacked server. TCP handshake process is exploited by the attacker sends half open connections seeking response from the server who's SYN-ACK never reach the destination. Servers keep the unestablished connections in queue for a period of time before discarding the packets (Linux OS leaves such connections open for 3 min each).
- UDP Flooding is initiated by excessively high volume of UDP datagrams IP network packets with MTUs ~1500 bytes being sent to random ports of the targeted servers or devices. The connectionless and non-mandatory packet transfer reliability feature of UDP packets make these fake packets unable to be reassembled. This causes the server resources to be consumed quickly which results in the target device being unavailable ultimately.
- ICMP Flooding is performed by redirecting ICMP echo requests to overload the target with requests which results in the server spending all its resources to respond to those requests and consumes the network bandwidth ultimately.

7.1.2.2 Application Layer DDoS Attacks

These attacks exploit application and server vulnerabilities by generating low-slow rate traffic, which looks legitimate and mimics human user behavior. These attacks overload the server and application resources and disrupt transmission of data between systems and hosts for the web application. These attacks are executed by introducing typical race conditions by requesting multiple computationally intensive GET/POST HTTP flood requests and monopolize transactions. This impacts the web portal performance, client reputation, and Quality of Service. XML and HTTP floods are some types of application layer attacks and mentioned below.

- HTTP flood targets web application servers and the web architecture flaws and vulnerability is using Slowloris and RUDY to send malformed HTTP packets in slow bandwidth traffic flow, sending partial requests, attempting to exceed the maximum concurrent connection pool that causes the web server to deny any more connection attempts from legitimate users or sending specialized HTTP GET or POST requests that exhaust the target server connection table.
- XML Flood employs the X-DoS markup language which is used for cloud communications with user and providers start with SOAP messages. These are written in XML and these user validation requests get exploited by simple XML tag changes. This allows unauthorized entry to the cloud services.

7.1.2.3 Reflection or Protocol DDoS Attacks

These attacks involve sending large number of requests similar to volumetric attacks which get amplified by redirecting from more such bots hosts using spoofed IP Address as shown by Arukonda and Sinha (2015). This leads to flooding of requests on the target, exhausting the connection state tables of the servers and the intermediate equipment by consuming the resources. In this attack, the request sent to a server has the response larger than that of the request. State exhaustion, SYN floods, DNS protocol flood, smurf attacks, and fragmented packet attacks are typical reflector DDoS attack examples and this is measured in millions of packets per seconds.

- State exhaustion DDoS has the attacker targeting the firewalls, routers, or server hosts to overwhelm and exhaust the maximum finite simultaneous connections which the server or network devices can support. These attacks target vulnerabilities in servers, operating systems, and holes in the network infrastructure to significantly impact the availability of some or all of the infrastructure devices and impact multiple tenants.
- Protocol DDoS Attacks: Targeting protocols like DNS or NTP is becoming a concern for cloud providers. DNS amplification can be done by targeting a misconfigured DNS server and sending a 64 byte while UDP dig request from a spoofed IP. The command returns 3–4 Kbyte (50 times the request). NTP amplification attacks also have UDP as the attack vector.

7.1.3 DDoS Attack Tools

This research required use of software tools in order to perform DDoS attacks. These tools are reconfigured for the attack purpose and are described in this section.

- Low Orbit Ion Canon or LOIC launches floods of increasing garbage requests of TCP, UDP, and HTTP packets to overwhelm the target web server and disrupt its services by a single attacker. Figure 7.4 illustrates LOIC attack targeting web

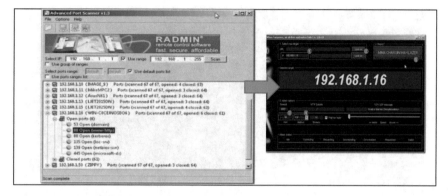

Fig. 7.4   LOIC Attack Process

server with user requests in low packet rate attacks as illustrated by Bhuyan et al. (2015).

- XOIC is designed to perform network DDoS attack on specific IPs, multiple URLs, user selected port, or user selected protocol. This software has the ability to cause HTTP flood with low number of bots and supports "booster files" to increase the magnitude of the attack, configurable VBScript modules to randomize HTTP headers of attacking systems as shown in Fig. 7.5 below.
- DDoSIM creates bots and zombies operating on Layer 7 with spoofed IP addresses as shown in Fig. 7.6 below with data packets having in-built TCP, UDP and HTTP, ICMP messages creating full TCP connection (SYN-SYN or ACK-ACK).
- Slowloris simulates an application layer attack by using slow read, HTTP POST and Apache range header attacks; these end up consuming significant memory and CPU of application servers. This runs on UNIX as shown in Fig. 7.7 below.
- Pyloris tool has ability to actually own HTTP request headers during an attack, keeping the connection open as long as required for exhausting the resources of the target server. Figure 7.8 below describes the attack parameters of the threads, requests, and time between threads.
- RUDeadYet or RUDY is a HTTP POST attack software which performs long field submission attack using POST method allowing the attacker to select forms and fields to use for the POST attack for connections to open with timeout and content length as shown in Fig. 7.9 below for a slow POST attack on a web site.
- Hulk or HTTP Unbreakable Load King generates unique requests to the target web servers. This attack software perform referrer forgery to bypass cache engines hitting the web server resource pool and avoid detection from IDS via known patterns as shown in Figure 7.10 below for requests sent and responses received.

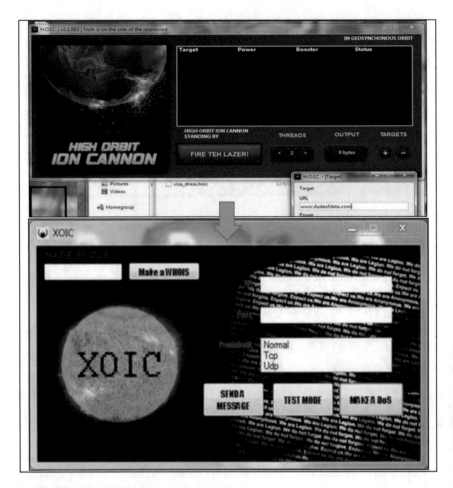

Fig. 7.5 XOIC Attack Process

7.2   Literature Review

However, the cloud computing market faces stability, efficiency and efficiency challenges, gaining recognition and deployment among organizations. Protection concerns for multinational entities and service providers are essential in the cloud and distributed application denials have the prime importance of all cloud-related risks. This chapter provides a study of studies in scholarly literature on cloud attacks by distributed denial of service and criteria for successful counter measurement strategies. This section reviews IEEE, ACM, Science Direct, and other digital libraries in the fields of cloud computing and DDoS attacks from January 2010 to December 2019. Research reported on the basis of cloud protection, DDoS prevention, DDoS identification, hybrid cloud, network design, packet flow, SYN, TCP and UDP flood is analyzed. While research work and literature surveys have already been submit-

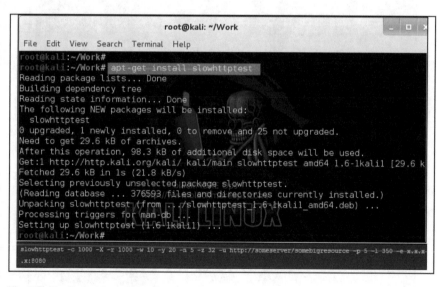

Fig. 7.6 DDoSIM Attack Process

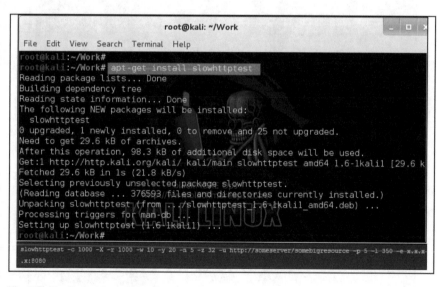

Fig. 7.7 Slowloris Attack Process

ted in the area of DDoS domain, this literature survey is different in the following ways.

- The study on DDoS attacks on cloud systems and networks was done by Wong and Tan (2014), while DDoS mitigation was the subject of this portion.
- Number of other studies and polls including Darwish et al. (2013) and Prabhadevi et al. (2014) are of small relevance.

Fig. 7.8 Pyloris Attack Process

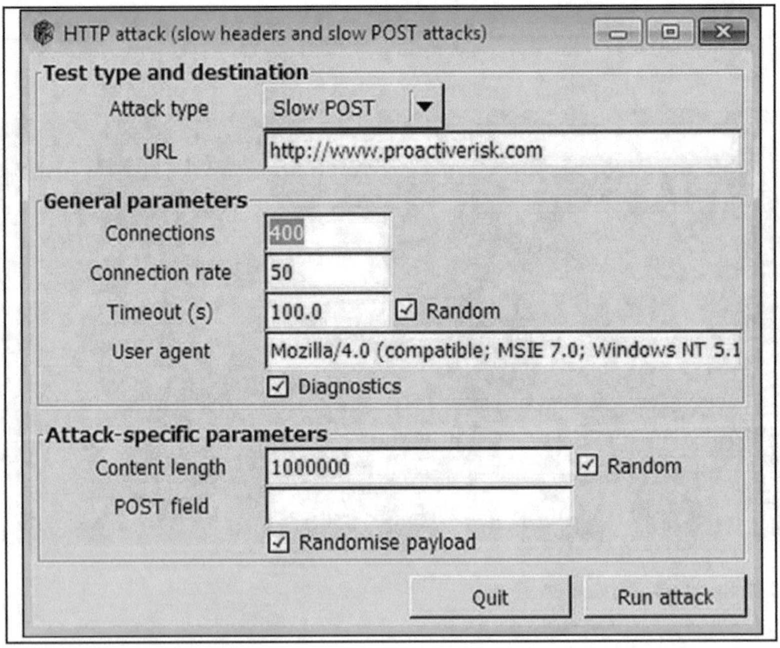

Fig. 7.9 RUDY Attack Process

- Consequences of DDoS cloud attacks are stressed by Malik and Mohammed (2012) in research articles based on hybrid clouds.
- Merlo et al. (2014) describe the DDoS attacks on cloud networks, while the focus of the study is on hybrid cloud architecture and design.

Fig. 7.10 HDoS Attack Process

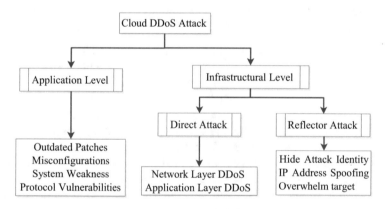

Fig. 7.11 DDoS attacks on clouds

The illustration in Fig. 7.11 above illustrates DDoS cyberattacks on cloud environments.

Table 7.1 presents the DDoS attack types, features, and tools.

The summary of cloud DDoS attack references as stated in the review literature is illustrated in Table 7.2 below for the research papers reviewed from the year 2010 to 2016 regarding the detection techniques, location, and the level of DDoS attack.

From the table summary, the most common deployment location for DDoS defense tends to be the access point while the DDoS attack level is primarily aimed at infrastructure levels.

Table 7.1 DDoS attack types and tools

Attack types	DDoS type and characteristics				Attack tools
	Infra	Application	Direct	Reflector	
ICMP Flood	√		√		LOIC
SYN Flood	√		√		TFN
UDP Flood	√	√	√	√	LOIC
HTTP Flood		√	√		DDoSIM
XML Flood		√	√		DAVOSET
Ping of Death	√		√		PING
Slowloris		√	√		Pyloris
Smurf	√			√	Nemesis

7.2.1 DDoS Attack Classification

In order to understand the DDoS attacks better, attack types are classified as per degree and level of attack automation, vulnerabilities exploited, attack rate dynamics, and attack impact as shown in Fig. 7.12 and presented in the section below.

- Manual attacks include a network search intruder, IP addresses, vulnerability computers, a device collapse, a firmware execution, and a malicious payload for remote users' access to the system as per degree of attack automation. Then the machine is able to initiate an attack against the attacker's order. Half-automatic attacks include inspecting the user computers, using attack commands, downloading payload, and inserting attack codes. The victim system is managed by the supervisors who select the form and the target victims when and how. Automatic attacks are carried out with a high automation degree that the infected user interfaces have a code for attack and pre-determined program attack type, length and IP address of the target. After the payload has been launched or automated threats, there is little engagement.
- Bandwidth depletion attacks include the overflowing of and amplifying the WAN pipelines with network assault packets. According to the vulnerability exploitation. The flood involves bots and zombies which send huge amounts of traffic to block the target bandwidth pipes and congest them. The victim's reaction is slowed down as flood demands are raised, the bandwidth tube becoming saturated and access to registered users is blocked. Enlargement attacks involve the bots and zombies sending messages through broadcast to the target's subnet. Attacks to resource destruction include malformed data packets, which have zombies send incorrect IP packets, maliciously crashing them and using a particular protocol to make them impossible for legal users, so that the perpetrator will absorb energy.
- During cyber attacks, variable DDoS attacks are most frequent, as per attack rate dynamics. Continuous intensity assaults are carried out without delay or with a decrease in the assault power. This easily leads to service interruption, but this

Table 7.2 Summary of DDoS attack mechanism

Year	Reference	Detection type	Deployed at	DDoS level
2010	Lo et al.	Signature	Access point	Infrastructure
	Bakshi et al.	Signature	Access point	Infrastructure
2011	Kim et al.	Hybrid	Access point	Infrastructure
	Kwon et al.	Anomaly	Access point	Not defined
	Gul et al.	Signature	Access point	Not defined
2012	Karnwal et al.	Signature	Distributed	Application
	Bedi et al.	Anomaly	Access point	Not defined
	Chatterjee et al.	Hybrid	Access point	Not defined
	Chonka et al.	Hybrid	Access point	Not defined
	Modi et al.	Hybrid	Access point	Not defined
2013	Lonea et al.	Anomaly	Access point	Infrastructure
	Karnwal et al.	Signature	Distributed	Application
	Gupta et al.	Signature	Access point	Infrastructure
	Modi et al.	Hybrid	Access point	Not defined
	Zakarya et al.	Anomaly	Access point	Not defined
	Huang et al.	Anomaly	Access point	Infrastructure
	Lonea et al.	Signature	Access point	Infrastructure
	Choi et al.	Anomaly	Access point	Infrastructure
	Ismail et al.	Anomaly	Access point	Not defined
	Dou et al.	Anomaly	Access point	Not defined
	Negi et al.	Anomaly	Access point	Not defined
	Jeyanthi et al.	Signature	Access point	Not defined
	Gupta et al.	Hybrid	Distributed	Not defined
2014	Zareapoor et al.	Signature	Access point	Infrastructure
	Vissers et al.	Signature	Access point	Infrastructure
	Choi et al.	Signature	Distributed	Infrastructure
	Iyengar et al.	Signature	Distributed	Not defined
	Michelin et al.	Signature	Access point	Application
	Teng et al.	Signature	Access point	Infrastructure
2015	Gamble et al.	Signature	Access point	Infrastructure
	Girma et al.	Signature	Distributed	Not defined
	Wang et al.	Signature	Access point	Not defined
	Marnerides et al.	Signature	Access point	Infrastructure
	Chen et al.	Signature	Access point	Application
2016	Seyyed et al.	Hybrid	Access point	Infrastructure
	Selvaraj et al.	Signature	Access point	Application
	Wang et al.	Hybrid	Access point	Infrastructure

assault is also observed. The attacks differ in their attack frequency and intensity such that detections that result from a rise in attack force or a fluctuating attack rate are deliberately avoided.

- According to the attack effect, two typical forms of attacks are destructive and degrading. The consequences of malicious attacks will be shutdown entirely

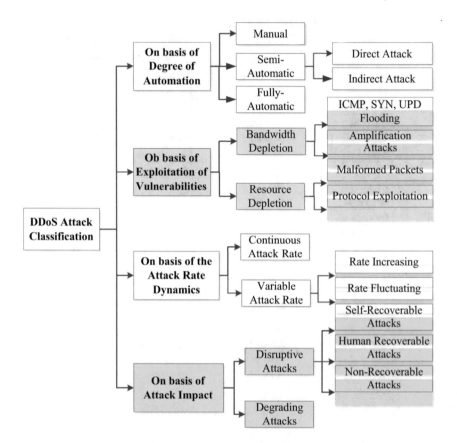

Fig. 7.12 DDoS attack classification

and the legal consumers will be refused facilities. Restoring from these malicious attacks has an effect dependent on artificial self-healing, human intervention or cannot be retrieved. Degrading assaults steadily exhaust the victim's resource in restricted spaces. This is far knowledgeable and harder to track than other threats.

Although some proposals for research and partial DDoS mitigation methods have been discussed above, most aim mainly to mitigate a few facets of the whole DDoS attack. No one appears to be fired at every known DoS attack by a full countermeasure. In an effort to counteract current and revolutionary steps, cyberattackers come forward regular with new vector challenges and attack derivatives. This leads to the fact that more research is needed to plan and create an efficient DDoS solution. Considering the parameters in the detection of DDoS attacks, the security mitigation mechanisms in Table 7.3 above are rated in scale 1 (low) to 10 (high).

Table 7.3 Comparing DDoS mitigation defense mechanisms

Attack detection parameters	Centralized			Distributed
	Source based	Destination	Network based	Hybrid
Accuracy	3	9	3	5
Scalability	3	4	6	6
Performance	6	7	6	5
Complexity	2	3	7	6
Overall defense	No	No	No	Yes

7.3 Methodology

The research adopted in this chapter is based on performing DDoS attacks after designing and implementing single-tier and three-tier architectures as well as on secondary DDoS survey. The research methodology phases are presented in Fig. 7.13 below.

Phase 1: Understand—Cybersecurity Survey and Literature Research

- Perform cybersecurity survey to understand the attack trends and impact.
- Review literature research on distributed denial of service, cloud computing and security algorithms to understand the security issues and vulnerabilities.

Phase 2: Analyze—Study existing cybersecurity DDoS mitigation solutions

- Review existing DDoS mitigation solutions and the current trend.
- Analyze existing mitigation solutions DDoS attacks on cloud computing environments.

Phase 3: Design and Implement—DDoS mitigation solution

- Design secure architecture to mitigate DDoS attacks on hybrid clouds.
- Perform DDoS attacks on proposed secure architecture and single-tier architecture.

Phase 4: Interpret—Validate results and conclusion

- Correlate results for the proposed secure architecture with single-tier design.
- Compare metrics and parameters for single-tier and three-tier architecture designs.
- Validate results and draw conclusions.

The broad scope of modern age dynamical vector-sensitive DDoS attacks can hardly be expected to take conventional IT security mechanisms such as on-site deployment, ISP data center facilities, or third-party scrubbing solutions. DDoS attacks are enough to block or absorb attacks by overwhelming the WAN circuit of a service provider. DDoS attacks not only interrupt functions, but distract intensity infrastructure, while other forms of attacks are cyberattacks attempted and prob-

Fig. 7.13 Research methodology

lems induced by system security weakness. With the volume and percentage of cloud service customers increasing the use of personal computers and mobile smartphones by home users and corporate staff. This has resulted in an uptick in cyberattacks on innocent people.

7.4 Review of Solutions for DDoS Attacks on Clouds

This section illustrates the current trends and threats posed by cyberattacks and presents results of the DDoS Survey conducted by the researcher for cyber-threats and attacks faced by organizations and cloud service providers. This section also reviews the DDoS mitigation strategies for different types of DDoS attacks and the existing DDoS solutions available for cloud.

7.4.1 Cyberattack Trends

The researcher initially reviewed cybersecurity attack and DDoS reports from Imperva and Akamai among other cybersecurity reports. Following are the primary trends seen for DDoS attacks.

- Latest ransomware and DDoS: Latest attacks by DDoS and ransomware are hard to protect, but malicious demands for DDoS vary only in their purpose to make legal requests, not their content.
- The hacker capabilities of large-scale attacks are gradually growing with a rise of 75% in peak attack (577 Gbps) relative to 2015, 274 attacks +100 Gbps and 47

attacks +200 Gbps were observed in 2016, as relative to 223 attacks in 2015 and 16 attacks overall in 2016.

- Large botnets advanced zombie attacks: 76% registered current BOT infections, heap-based buffer overflow vulnerability vectors on Linux servers, Microsoft SQL Reflection technique, and 49% cloud-based SaaS industries are attacked.
- Attack frequency: The United States, the UK, and France have been the top targets for attacks over 10 Gbps over the last 24 months with an average of 125,000 incidents each week.
- Impact: The amount of workers that have visited suspicious pages was 85%, the number of people who have downloaded a malware payload file was 94%, while the loss of information was 88%, and company losses 400%.

7.4.2   Cybersecurity Survey

An electronic survey was conducted with the focus on cyberattack threats and the impact on organizations. Using Survey Monkey as the contact medium questionnaire requests were sent to 700 IT security and industry professionals with responses received from 550 participants. In order to ensure the DDoS Survey had the right mix of target audience:

- The researcher ensured the industry representation of respondents involved Information Technology professionals from Cloud Computing, Information Security, Data Center, and Infrastructure Operations domains.
- The researcher also ensured the respondents belonged to a broad range of industries across different organizations with more than 1000 employees.
- The survey requests sent were evenly divided among domestic and international respondents with global locations and businesses utilizing cloud computing.

Detailed breakup, roles, and responsibilities are presented in the Tables 7.4 and 7.5 below.

The respondent's views and results obtained from the DDoS survey are as illustrated below. These are collected based on ten specific questions on mitigation strategies, responsibilities, impact, ability and barriers regarding distributed denial of

Table 7.4 Respondents roles

Respondent's roles	Breakup
Information security	33%
Security operations	18%
Network security	12%
IT support	09%
Systems admin	08%
Audit compliance	07%
Web deployment	06%
Data center ops	07%

Table 7.5 Respondent organizations

Organization	Count	Breakup
Financial services	66	12%
Education	05	01%
Information technology	245	45%
Retail, E-commerce	45	08%
Internet service providers	44	08%
Gaming	105	19%
Media and Travel	30	05%
Pharmacy	10	02%

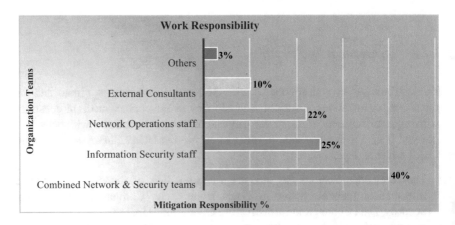

Fig. 7.14 Respondents work responsibilities

service attacks on the data centers and applications hosted on real-time production environments.

Survey Question #1: Responsibility in case of DoS attacks is recognized to perform incident response and mitigation?

While there is no one team dedicated to mitigate cyberattacks, most organizations deploy a team from the Network and Information Security domains from within the organization to work together with shared responsibility during a DDoS attack till the DDoS has been mitigated; in other words, a combined team drives the mitigation plan, as illustrated in Fig. 7.14.

Survey Question #2: Factors to determine impact of DoS attacks against the organization?

Commercial impact and high cost of technical repairs and support involved are the top two issues concerning organizations regarding the impact of DDoS attacks as illustrated in Fig. 7.15.

Survey Question #3: Enterprise has ability to detect and prevent DoS attacks?

Ideally every organization should have a mitigation plan ready but only 54% organization felt confident enough to confirm the ability to block and contain a

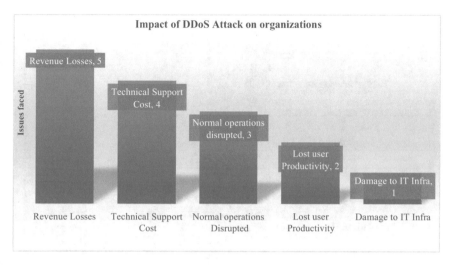

Fig. 7.15 Impact of DDoS attacks on organizations

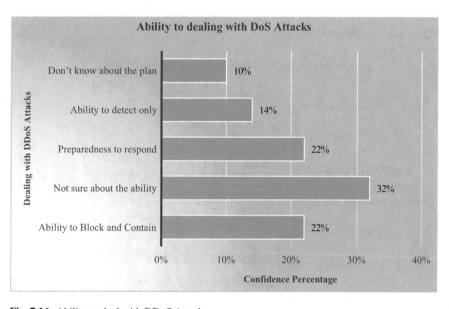

Fig. 7.16 Ability to deal with DDoS Attacks

DDoS attack. This is presented in Fig. 7.16 above. This assumption however is largely untested as most organizations only assume their ability and don't have a valid test result to prove the plan.

Survey Question #4: Categorize and rate impact areas due to DDoS attack.

Loss of trust with lower customer confidence is one of the most damaging consequences of DDoS attacks as the business takes a huge hit as presented in Fig. 7.17.

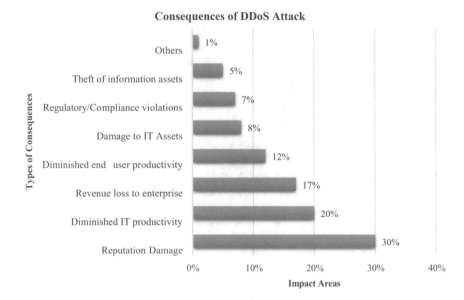

Fig. 7.17 Consequences of DDoS Attacks

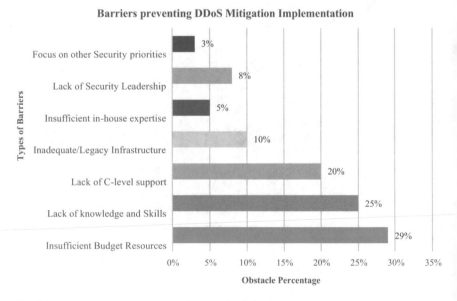

Fig. 7.18 Barriers preventing DDoS mitigation implementation

Survey Question #5: What are the barriers that prevent DDoS mitigation implementation?

Lack of budget and knowledge skills are the top obstacles which prevent DDoS mitigation implementations for the organizations as shown in Fig. 7.18.

Survey Question #6: Attacks resulting in maximum downtime for cloud hosting services?

DDoS and malware are the top ranked cyber-threats for most organizations worldwide as illustrated in Fig. 7.19 below.

Survey Question #7: DoS attacks caused downtime of the data center?

Unplanned data center outages primarily due to DDoS resulted in 45% respondents confirming entire data center operations were shut down as illustrated in Fig. 7.20.

Survey Question #8: Enterprise security resilience against DDoS attacks validated?

Majority of organizations (72%) recognize the need to have the security assessment at least once, while 49% performed the checks more than once annually. This overall is a good trend as it points to organizations taking notice of the cyberattacks and seeking to be prepared against them as presented in Fig. 7.21.

Survey Question # 9: Mitigation capabilities activated during DDoS attacks?

Organizations still rely on service providers to block DDoS on WAN circuits or the on-premise deployments during DDoS attacks as presented in Fig. 7.22 below.

Survey Question #10: Most important factors during DoS attacks?

Ability to detect DDoS attacks with as little human intervention and the reporting around the cyberattack with visibility are the most critical factors which an organization needs to have in place as illustrated in Fig. 7.23.

The aforementioned survey findings show that security decisions are driven by security management and most companies have little or no knowledge of the real impact of DDoS attacks. The most damaging effect is the loss of customer trust and confidence. This adds to more setbacks and problems with technological assistance.

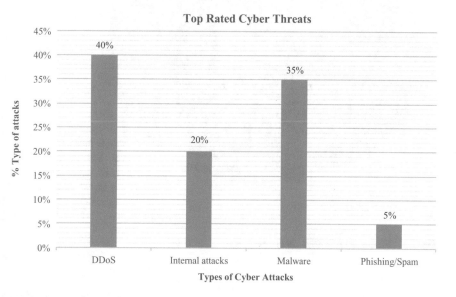

Fig. 7.19 Top rated cyber-threats

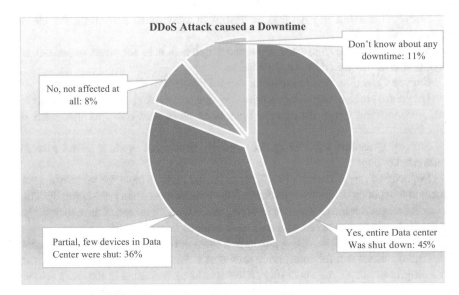

Fig. 7.20 Downtime resulted due to DDoS Attacks

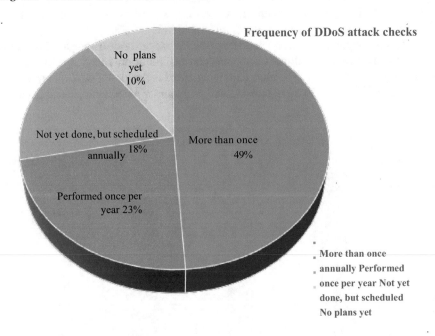

Fig. 7.21 Frequency of DDoS checks performed

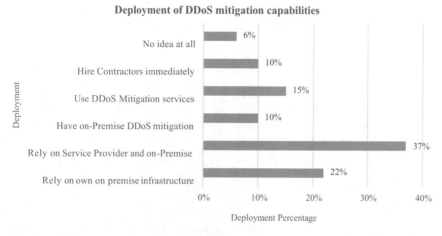

Fig. 7.22 Mitigation activities during DDoS attacks

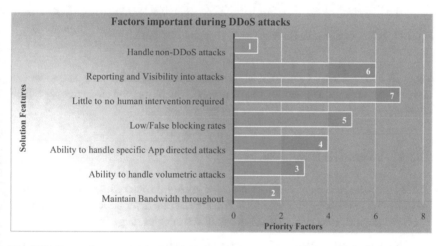

Fig. 7.23 Factors important during DDoS attacks

Currently, only 55% said they had ample faith in stopping cyberattacks. Total DDoS attacks and ransomware are the highest rated cyber-threats among all industry-wide organizations.

7.4.3 DDoS Mitigation Strategies

DDoS mitigation needs to be seamless and comprehensive in order to protect the cloud services and web hosting against the DDoS attacks which are aimed at different layers of the TCP stack as described by Khadke et al. (2016) and illustrated in

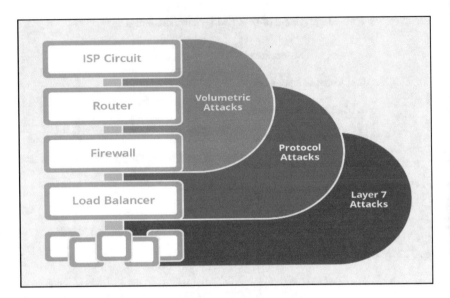

Fig. 7.24 DDoS attacks on data center devices (Imperva 2016 Report)

Fig. 7.24 above. There is a need to address each type of DDoS with a unique toolset and defense strategy. The section below presents the mitigation strategies for DDoS attacks.

7.4.3.1 Volumetric DDoS Attacks

Volumetric attacks typically impact network layer 4 devices while protocol attacks occur on load balancers and firewalls. The application layer attacks occur on layer 7 server systems. Volumetric DDoS mitigation strategies range from:

- Block with On-Premise Devices: Attempting to block a volumetric DDoS attack using on-premise devices such as IPS/IDS and firewalls are typically ineffective. As these devices are positioned in the network downstream from the point at which the DDoS traffic causes saturation of the link and packet loss.
- Null Routing the Target IP: An organization using Border Gateway Protocol (BGP) may use null routes to prevent devices on the Internet from sending traffic to the target IP. The main benefit of this approach is its distributed effect, as the null route announcement is sent to all devices on the Internet that receive Internet routing table announcements. Since this mechanism results in the target IP becoming unreachable, it is only useful if the target IP is expendable and traffic can be discarded to save other resources on the same network. For this reason, null routes are typically only an effective mitigation in multi-user environments where a problematic user can be segmented off to ensure availability for the remaining users.

- Hide Behind a Large Content Distribution Networks: Traditional CDNs function by locating web server caches throughout the world to deliver content to the Internet. A CDN typically works as an HTTP(S) proxy where all requests are made to the CDN server, which subsequently initiates a private connection to the organization's server to obtain the data. Volumetric flood attacks using CDNs often implicitly protect a network from these types of attacks because the traffic is sent to the CDNs which often is comprised of a massive, globally distributed network.
- Blocking Upstream by the ISP: If contacted, the technical support arm of most ISPs will add simple rules to block specific traffic before it reaches the target network. The limitation is related to the minimal filtering capabilities offered by most ISPs. For example, as illustrated in Fig. 7.25 below, most ISPs filter traffic to or from a specific IP address, or using a certain protocol. But this is a very crude method, which may not be granular enough to block attack traffic and allow the legitimate traffic.
- Dedicated Mitigation Services: A dedicated DDoS mitigation service is often the most effective approach. These services are similar to the CDN approach as mentioned earlier to identify and block DDoS traffic. Much like CDN, reputable mitigation services have a massive globally distributed network of scrubbing centers capable of blocking large DDoS attacks.

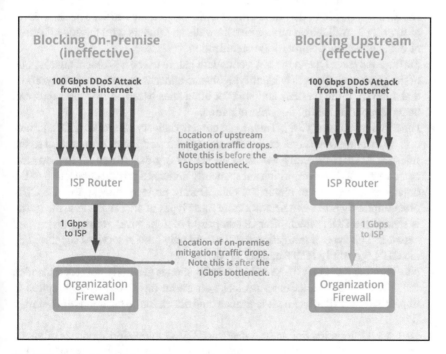

Fig. 7.25 Traffic blocking by ISP (Imperva 2016 Report)

7.4.3.2   Protocol DDoS Attacks

Unlike volumetric attacks, the protocol DDoS attacks intention is not to saturate the Internet connection but to cause disruption with a relatively small amount of network traffic. Protocol DDoS mitigation strategies range from the following.

- Blocking with On-Premise Devices: Blocking protocol DDoS attacks with on-premise devices such as IDS/IPS and firewalls may be successful due to the low bandwidth nature of these attacks. This is in contrast to volumetric DDoS attacks where the bottleneck is upstream and outside the control of the organization. Also, unlike volumetric attacks, a large portion of protocol attacks do not have spoofed source IP addresses. As a result, simple attacks can be blocked with simple firewall rules. More advanced attacks, particularly those sourced from very large botnets, will require purpose-built DDoS mitigation hardware to properly identify and automatically block the attack traffic.
- Blocking Upstream by the ISP: This method of mitigation is often ineffective for protocol DDoS attacks. If contacted, the technical support arm of most ISPs will add simple rules to block specific traffic before it reaches the target network. This approach can be effective at mitigating simplistic attacks, but will often be unable to mitigate more complex scenarios. The limitation is related to the minimal filtering capabilities offered by most ISPs. For example, most ISPs will be happy to filter all traffic to or from a specific IP address, or all traffic using a certain protocol. But this is a very blunt tool and may not be granular enough to block DDoS traffic while at the same time allowing legitimate traffic. The same level of filtering is available in on-premise firewalls and routers with the added benefit of being under the control of the organization.
- Traffic Analytics: Due to the low bandwidth nature of most protocol attacks, one of the greatest challenges is identifying that an attack is actually underway. Tools that analyze traffic patterns and look for anomalies based on historical data can be invaluable in making this determination.
- Hide Behind a Large CDN: Traditional content delivery networks (CDNs) function by locating web server caches throughout the world to deliver content to the Internet. A CDN typically works as an HTTP(s) proxy where all requests are made to the CDN server, which subsequently initiates a private connection to the organization's server to obtain the data. During protocol attacks using a CDN often implicitly protects a network from these types of attacks because the traffic is sent to the CDN which often is comprised of a massive, globally distributed network. This type of mitigation solution will only protect services supported by the CDN (generally HTTP and HTTPS).
- Null Routing Target IP: This method of mitigation is generally not recommended for protocol DDoS attacks as it blocks all traffic to the target. This option is imprecise and will affect both legitimate and attack traffic destined for the target IP.
- Dedicated Mitigation Services: A dedicated DDoS mitigation service is often the most effective. These services are similar to the CDN approach mentioned earlier

but with advanced capabilities specific to identifying and blocking DDoS traffic. Much like a CDN, a reputable mitigation service has a massive globally distributed network of scrubbing centers capable of blocking large attacks.

7.4.3.3 Application Layer DDoS Attacks

It's critical that the attack traffic is "in protocol"; this makes traffic legal from a protocol viewpoint. The primary difference between application-level and other attacks. The attacks are also difficult, as defined by Durcekova et al. (2012), to discern from legitimate traffic. Application DDoS attack mitigation strategies range from the following:

- Blocking with On-Premise Devices: Blocking application-level DDoS attacks with on-premise devices such as IDP/IPS and firewalls may be successful due to the low bandwidth nature of these attacks. This is in contrast to volumetric DDoS attacks where the bottleneck is upstream and outside the control of the organization. Unlike volumetric or protocol attacks, nearly all (TCP specifically) application-level attacks do not have spoofed source IP addresses. As a result, simple attacks can be blocked with simple firewall rules. More advanced attacks, and those sourced from very large botnets, will require purpose-built DDoS mitigation hardware to properly identify and automatically block the attack traffic.
- Blocking Upstream by the ISP: This method of mitigation is often ineffective for application-level DDoS attacks. If contacted, the technical support arm of most ISPs will add simple rules to block specific traffic before it reaches the target network. This approach can be effective at mitigating simplistic attacks, but will often be unable to mitigate more complex scenarios. The limitation is related to the minimal filtering capabilities offered by most ISPs. For example, most ISPs will be happy to filter all traffic to or from a specific IP address, or all traffic using a certain protocol. But this is a very blunt tool and may not be granular enough to block DDoS traffic while at the same time allowing legitimate traffic. The same level of filtering is available in on-premise firewalls and routers with the added benefit of being under the control of the organization. In addition, unless they are given decryption keys, the ISP is unable to inspect the content of traffic using encrypted protocols like HTTPS, making identification and mitigation more difficult.
- Traffic Analytics: With low bandwidth application attacks, one of the greatest challenges is identifying that an attack is actually occurring. Tools that analyze traffic patterns and look for anomalies based on historical data can be invaluable in making this determination.
- Null Routing the Target IP: This method of mitigation is generally not recommended for application-level DDoS attacks as it blocks all traffic to the target. This option is imprecise and will affect both legitimate and attack traffic destined for the target IP.

- Hide Behind a Large CDN: Traditional content delivery networks (CDNs) function by locating web server caches throughout the world to deliver content to the Internet. A CDN typically works as an HTTP(s) proxy where all requests are made to the CDN server, which subsequently initiates a private connection to the organization's server to obtain the data.
- Regarding application-level attacks, using a CDN may help mitigate some attacks. Specifically, requests for resources located on the CDN, such as static web objects, will be fulfilled and absorbed by the massive CDN infrastructure.
- However, dynamic content such as user login requests, content searches or similar non-cacheable data, will be passed by the CDN to the organization's backend servers resulting in a DDoS attack. Moreover, this type of mitigation solution will only protect services supported by the CDN (generally HTTP and HTTPS).
- Dedicated Mitigation Services: A dedicated DDoS mitigation service is often the most effective approach to solving application-level attacks. These services are similar to the CDN approach mentioned earlier. Much like a CDN, a reputable mitigation service has a massive globally distributed network of scrubbing centers capable of blocking large DDoS attacks.
- Application Blocking: For smaller application-level attacks, an organization may be able to mitigate the attack by disabling the feature being targeted. For example, if the attacker is targeting a search feature on the site that is inefficient, it may be better to temporarily disable that feature to maintain a proper level of performance for the other components on-site.

7.4.3.4 Reflection Attacks

Reflection attacks are amplification attacks similar to volumetric DDoS attacks using the same protocol in both directions. Server responses sent to the source are substantially greater than the requests.

Attackers take advantage of such a scenario and direct the response traffic by magnifying and flooding the victim with unwanted traffic that overwhelms the network circuits and servers. The notion of amplification is presented by Georgios et al. (2007) for the DNS reflection attack as illustrated in Fig. 7.26 below. 72 byte queries like RRSIG and DNSKEY to an open vulnerable DNS resolver server results in 112 byte response. ICMP smurf attacks on publically accessible UDP systems are examples of such attacks. Attackers send the spoofed request (64 byte) which reaches an open resolver server and is then reflected to the victim as 3876 bytes. The mitigation option is to use BCP38 which allows network routers to validate IP addresses in case the attackers try to spoof the IP address, the DNS server can block reflecting traffic to the victim and only packets originating from valid IP addresses would be replied back.

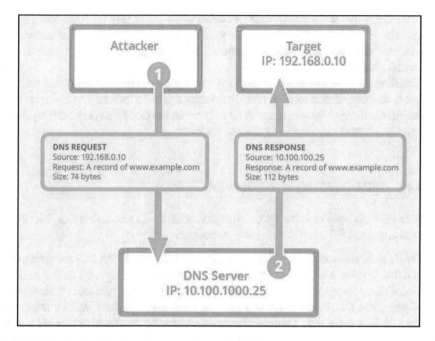

Fig. 7.26 Reflection DDoS attack process (Imperva 2016 Report)

7.4.4   Review of DDoS Mitigation Solutions

The biggest threat, multiplicities by the sophistication, volumetric storms, business businesses, bankers, bank hosting firms, and financial and network firms, has been a distributed denial of service assault on cloud providers. DoS attacks have been key. In all, most DDoS mitigation solutions as shown by Sridaran and Nagaraju (2016) are unable to provide a proper and adequate protection against varied levels of network or application attacks, and always seem to lack the features to mitigate and block the new types of attacks that are constantly evolving. To provide a solid DDoS protection, a robust, secure and scalable solution is required. This section presents the traditional solutions for mitigating DDoS attacks.

7.4.4.1   On-Premise DDoS Mitigation Solutions

On-premise infrastructure such as a private cloud with limited ISP leased bandwidth, basic security devices such as firewalls and IDS, as proposed by Hildmann and Odej (2014). Even though an in-house on-premise defense system may have DDoS mitigation defense functionalities, however it would not be able to truly deliver a proper DDoS mitigation due to the following:

- The inability of in-house defense systems in defending against volumetric floods—when attacks flood and saturate the ISP WAN circuits and the enterprise defense network themselves—becomes a challenge to stop high-volumetric attacks on the networks.
- Another issue is the constant need for an ongoing investment on IT infrastructure, training, and resources with increasing dynamic threats. Most enterprises using cloud services would not want to have an internal IT or dedicated security groups or invest additional redundant resources.

7.4.4.2   ISP DDoS Mitigation Solutions

While ISPs do tend to offer DDOS mitigation as an additional service, blocking DDoS attacks at ISP level does have drawbacks.

- With multiple customers sharing the same WAN link and the ISP providing the DDoS Service solutions using common equipment during an attack, the ISP would face issues with Internet traffic for each and every "protected" customer. During the DDoS attack on one specific customer, the ISPs WAN equipment would be galvanized to handle the increased traffic flood which would in turn affect other customers who are not targeted.
- Having multiple customers with hundreds of policies to implement like blocking IP addresses, black listing domains, allow/deny ports to avoid any false positives, ISPs would at times lower their guard by "softening" their policies and lower the alert thresholds. This can result in some malicious traffic getting passed through which even if is not a flood attack; it could lead to application attack. At times, the attacker traffic ends up behaving in a similar manner to a legitimate user's traffic request, thus leading to the ISP not being able to protect against dual network and application DDoS attack.
- ISPs core business area is network data delivery and is focused on providing WAN circuit uptimes and load balancing, expecting decent DDoS expertise would be asking a lot from vendors. The cost consideration for having multiple ISPs who may have implemented BGP or WAN load balancing circuits for which implementing a DDoS protection service would require additional services to be taken from each WAN provider.

7.4.4.3   Scrubbing Defense DDoS Mitigation Solutions

Use of scrubbing defense architecture is performed in two ways for DDoS protection as described by Zilberman et al. (2015). Either ways have all the traffic go through a third-party defense systems and send the cleaned traffic to the customer's network OR use two detection systems, one placed in house or on the data center premise at network perimeter level and the second mitigation system based at the

Security Operations Center (SOC) at the cloud data center level. These defenses complement each other in providing a quick and early detection for the attack types.

- The defense system at customer premise performs traffic analysis, attack detection, and signaling by constantly monitoring network traffic and the traffic pattern in order to establish a normal behavior baseline threshold much like an IDS. Then detect anomalies and denial attacks at initial stage and instantly alert the Data Center Security Operation Center for mitigation.
- If a volumetric refusing attack is carried out on the WAN circuit network, customer traffic is sent to the data center for scrubbing to obstruct and relieve flow. The scrubbed traffic is passed back to the cloud provider of the abonor until the initial filtering is completed. Scrubbing center teams gathered and processed attack data to allow for real-time and predictive analysis.
- There are however issues of compliance and regulations, the need to install detection systems as either a hardware device or a thick client for each customer and data privacy issues for traffic flowing to a third-party scrubbing center.

The DDoS attack mitigation methodology followed in this for executing the DDoS on the designed architecture is presented in detail in further sections.

7.5   Experimental Results

This investigates results obtained after performing the network- and application-level DDoS attacks on the single-tier and three-tier infrastructures. The criteria for analyzing the success and failure of the architectures is based on real user monitoring parameters such as ICMP response, browser throughput, page load response, and application server response.

7.5.1   Performance Results: Single-Tier Architecture

Real user monitoring parameter values for ICMP, page load, browser throughput, and application server response obtained before and during the DDoS attacks on single-tier architecture are presented in Fig. 7.27 below.

7.5.2   Performance Results: Three-Tier Architecture

Real user monitoring parameter values for ICMP, page load, browser throughput, and application server response obtained before and during the DDoS attacks on the three-tier architecture are presented in Fig. 7.28 below.

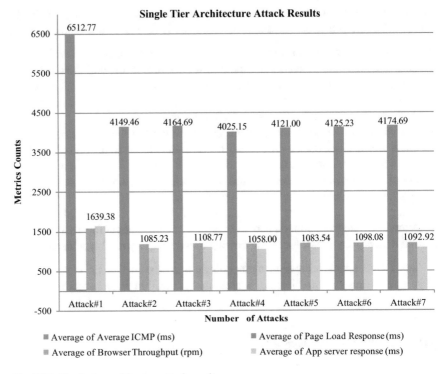

Fig. 7.27 Single-tier architecture attack results

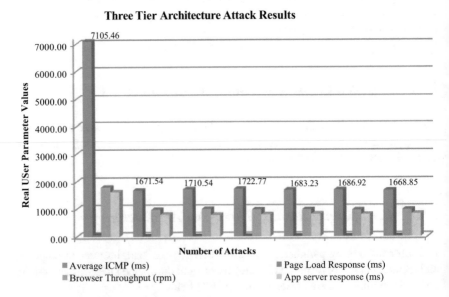

Fig. 7.28 Three-tier architecture attack results

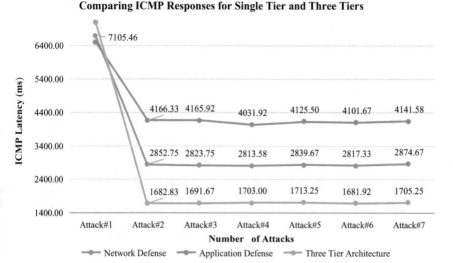

Fig. 7.29 ICMP response comparison results

Comparing single-tier and three-tier architecture results for ICMP responses during and after the DDoS attack are presented in Fig. 7.29.

Comparing single-tier and three-tier architecture results for page load responses during and after the DDoS attack are presented in Fig. 7.30.

Comparing single-tier and three-tier architecture results for browser throughput during and after the DDoS attack are presented in Fig. 7.31.

Comparing single-tier and three-tier architecture results for ICMP responses during and after the DDoS attack are presented in Fig. 7.32.

7.6   Designing and Implementing Architectures

7.6.1   Single-Tier Architecture

The researcher designed and implemented the single-tier architecture as a flat single-tier architecture with standard network services and web application portal simulating the cloud hosted application in a data center. The web portal comprises of web pages running scripts gathering real-time data like temperatures, NSE stock values and saving them on a database. This simulated the web portal application.

The single-tier architecture is designed with the standard routing and switching network devices running in an on-premise private data center, connecting the web portal to the Internet (Joshi et al. 2012) as illustrated in Fig. 7.33 below. The red

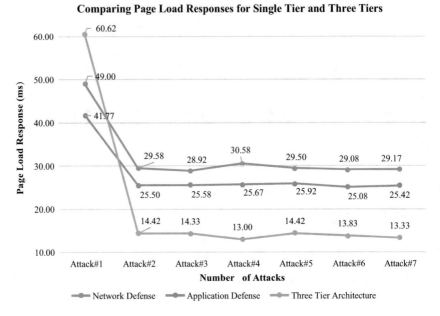

Fig. 7.30 Page load response results

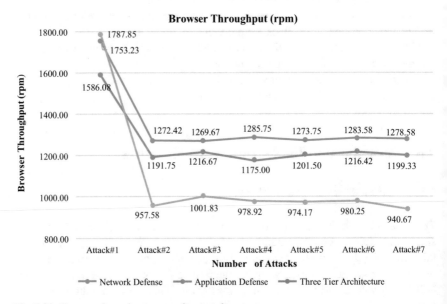

Fig. 7.31 Browser throughput comparison results

arrows denote the attack traffic, blue arrows designate the user traffic while the green arrows illustrate the outbound traffic. This design has the same single inbound and exit default gateway (11.252.15.1) for the web traffic. The legitimate users as well as attackers enter and exit following the same data flow route. The web

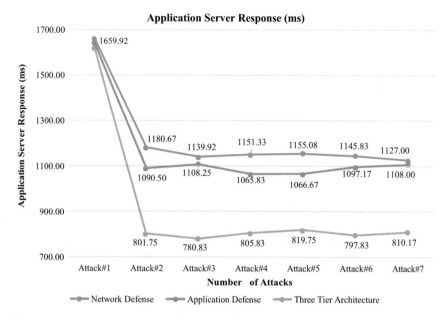

Fig. 7.32 Real user monitoring—Application server response

application running .NET and IIS services comprises of VMware virtual machines hosting the front end web portal (11.252.15.200) and backend SQL Server Database (11.252.15.25) which simulates the cloud application environment.

Hardware devices and servers implemented for the single-tier data center infrastructure and DDoS attacks are executed on the single-tier infrastructure as ICMP flooding with 1000 echo requests with increasing buffer size (from 3700 to 3805 bytes) as

Ping –t –l 202.122.134.4

Application-level attacks are executed using tools such as LOIC, R.U.D.Y, and Slowloris. Slow socket buildup is performed for slow web attacks and HTTP floods by increasing the thread count as

GET   /app/?id=436753msg=BOOM%2520HEADSHOT!HTTP/1.1Host: 11.152.15.201

These attacks simulated DDoS network- and application-level attacks that deny legitimate users the access to the web application portal running on the single-tier

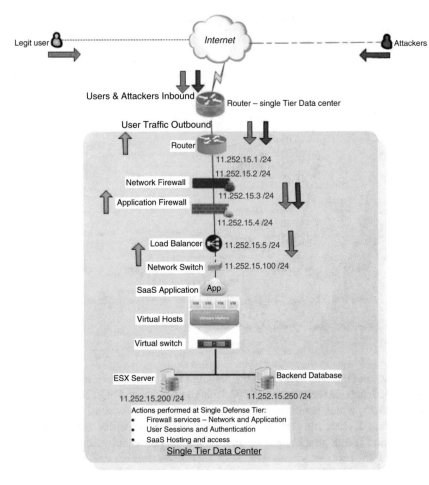

Fig. 7.33 Single-tier architecture design

infrastructure. Logs are collected from the network devices and servers during the DDoS attacks for detailed analysis.

Figure 7.34 below presents the raw logs obtained from the single-tier data center network device after the DDoS attack. This shows the excess DDoS HTTP rate and large-sized data packets depicting the DDoS attack. After detecting the DDoS attack, network firewall defense is initiated with logs are collected as illustrated in Fig. 7.35 below.

From initial analysis of these system and device logs, the single tier displays a steadily degraded performance and response towards the end user accessing the web portal. To further validate this theory, the single tier logs are analyzed for real user monitoring parameters and is investigated further and presented in detail in the subsequent chapter.

```
Raw Logs: Attack 1
Jan 28 2016 13:00:07 Warning [DDOS]:791366: UDP packet rate exceeded. Flow 192.168.0.100:2435 -> 11.252.15.100:2000. Limit 30. Current 3.
Jun 27 2016 13:30:15 Warning [DDOS]:810166: UDP packet rate exceeded. Flow 192.168.0.100:2435 -> 11.252.15.100:2000. Limit 30. Current 3.
Jan 27 2016 14:00:29 Warning [DDOS]:708372: DDoS packet L4 payload size is too big. Flow 226.61.80.115:53 -> 11.252.15.100:4696. Maximum 1280. Current 690.
Jan 27 2016 14:30:49 Warning [DDOS]:358374: DDoS packet L4 payload size is too big. Flow 158.91.47.243:53 -> 11.252.15.100:2001. Maximum 1280. Current 1.
Jan 27 2016 15:00:29 Warning [DDOS]:698373: DDoS packet from well-known UDP source port on 11.252.15.100 port 4619 has been detected. Current 9514.
Jan 27 2016 15:30:11 Warning [DDOS]:687298: DDoS HTTP destination request rate exceeded. Flow 192.168.0.100:42091 -> 11.252.15.100:80. Limit 33. Current 870.
Jan 27 2016 16:00:15 Warning [DDOS]:635606: UDP packet rate exceeded. Flow 192.168.0.100:2435 -> 11.252.15.100:2000. Limit 30. Current 3.
Jan 27 2016 16:30:29 Warning [DDOS]:708372: DDoS packet L4 payload size is too big. Flow 226.61.80.115:53 -> 11.252.15.100:4696. Maximum 1280. Current 861.
Jan 27 2016 17:00:44 Warning [DDOS]:358374: DDoS packet L4 payload size is too big. Flow 158.91.47.243:53 -> 11.252.15.100:2001. Maximum 1280. Current 1.
Jan 27 2016 17:30:19 Warning [DDOS]:698373: DDoS packet from well-known UDP source port on 11.252.15.100 port 4619 has been detected. Current 8954.
Jan 27 2016 18:00:17 Warning [DDOS]:628372: DDoS HTTP destination request rate exceeded. Flow 192.168.0.100:42091 -> 11.252.15.100:80. Limit 30. Current 693.
Jan 27 2016 18:30:17 Warning [DDOS]:638492: DDoS HTTP destination request rate exceeded. Flow 192.168.0.100:49031 -> 11.252.15.100:80. Limit 35. Current 547.
Jan 27 2016 19:00:09 Warning [DDOS]:793699: UDP packet rate exceeded. Flow 192.168.0.100:2435 -> 11.252.15.100:2000. Limit 30. Current 3.


Raw Logs: Attack 2
Jan 28 2016 13:00:07 Warning [DDOS]:4827475559: UDP packet rate exceeded. Flow 192.168.0.100:2435 -> 11.252.15.100:2000. Limit 30. Current 2.
Jan 28 2016 13:30:15 Warning [DDOS]:3409294068: UDP packet rate exceeded. Flow 192.168.0.100:2435 -> 11.252.15.100:2000. Limit 30. Current 1.
Jan 28 2016 14:00:29 Warning [DDOS]:5638205: DDoS packet L4 payload size is too big. Flow 226.61.80.115:53 -> 11.252.15.100:4696. Maximum 1280. Current 3498.
Jan 28 2016 14:30:49 Warning [DDOS]:5833412: DDoS packet L4 payload size is too big. Flow 158.91.47.243:53 -> 11.252.15.100:2001. Maximum 1280. Current 2.
Jan 28 2016 15:00:29 Warning [DDOS]:4373407: DDoS packet from well-known UDP source port on 11.252.15.100 port 4619 has been detected. Current 8954.
Jan 28 2016 15:30:11 Warning [DDOS]:8362128: DDoS HTTP destination request rate exceeded. Flow 192.168.0.100:42091 -> 11.252.15.100:80. Limit 30. Current 839.
Jan 28 2016 16:00:15 Warning [DDOS]:3691013566: UDP packet rate exceeded. Flow 192.168.0.100:2435 -> 11.252.15.100:2000. Limit 30. Current 3.
Jan 28 2016 16:30:29 Warning [DDOS]:7083702: DDoS packet L4 payload size is too big. Flow 226.61.80.115:53 -> 11.252.15.100:4696. Maximum 1280. Current 2696.
```

Fig. 7.34   Raw DDoS attack log

Attack#	Time (pm)	Buffer Size (bytes)	Echo Requests	Target Server IP	Real User Monitoring					Attack Vector Details
					Average ICMP (ms)	Page Load Response (ms)	Browser Throughput (rpm)	App server response (ms)	Status code	
Attack#1	13:00	3700	1000	11.252.15.100	6545	45	1800	1636	200	No standard network layer defense in place - single tier architecture Ping AppServer -n 1000 -l 3xxx Size: 3xxx, Echo request count: 1000
	13:30	3750	1000	11.252.15.100	6670	54	1856	1496	429	
	14:00	3760	1000	11.252.15.100	6575	55	1727	1624	200	
	14:30	3780	1000	11.252.15.100	6791	46	1627	1784	200	
	15:00	3790	1000	11.252.15.100	6583	41	1606	1713	429	
	15:30	3795	1000	11.252.15.100	6745	55	1806	1686	204	
	16:00	3800	1000	11.252.15.100	6790	50	1651	1488	429	
	16:30	3820	1000	11.252.15.100	6794	54	1761	1795	204	
	17:00	3810	1000	11.252.15.100	6690	47	1800	1833	503	
	17:30	3805	1000	11.252.15.100	6512	42	1849	1565	503	
	18:00	3820	1000	11.252.15.100	6692	48	1835	1726	503	
	18:30	3810	1000	11.252.15.100	6589	50	1635	1570	503	
	19:00	3805	1000	11.252.15.100	6995	50	1839	1663	503	
Attack#2	13:00	3750	1000	11.252.15.100	2795	30	1325	1297	200	Network Firewall Defense implemented: Attack vector categories of attack as ICMP/UDP/SYN floods
	13:30	3745	1000	11.252.15.100	2911	32	1327	1243	200	
	14:00	3760	1000	11.252.15.100	2805	29	1208	1298	200	
	14:30	3780	1000	11.252.15.100	2963	30	1306	1043	200	
	15:00	3770	1000	11.252.15.100	2746	29	1235	1097	200	
	15:30	3783	1000	11.252.15.100	2933	32	1245	1213	200	
	16:00	3780	1000	11.252.15.100	2988	28	1219	1228	200	

Fig. 7.35   Single-tier real user monitoring parameters

7.6.2   Three-Tier Architecture

Three-tier architecture is designed to mitigate DDoS attacks, Fig. 7.36 illustrates the tier-based design model with services running at each tier and mitigation performed.

The proposed infrastructure has three tiers, the first and second tiers are implemented as defense tiers against the DDoS attacks while the third tier is hosting the actual cloud portal and is the access tier. The three tiers are interconnected with each other via a secure virtual private network. Figure 7.37 below illustrates the network diagram in detail regarding the IP Address scheme, services at each level, and the device details.

- **Network Defense**
- Layer 3-4 Network devices
- Network Firewall, Tier1 Load Balancing, IP Reputation Blacklisting, User Authentication

- **Mitigates:**
- SYN, ICMP, TCP Floods, Malformed Packets

Public Cloud Tier 1

Public Cloud Tier 2

- **Application defense**
- Layer 7 devices

- SSL Termination, Web App Firewall, Secondary Load Balancing, User Session, DNS, Outbound traffic

- **Mitigates:** Slow POST, UDP floods

- **Access & Hosting**

- Final Authentication
- Access to Tier 3 using jump box server only
- No direct traffic flow

- Web Servers
- Application systems
- Backend Database

Private Cloud Tier 3

Fig. 7.36 Three-tier architecture design model

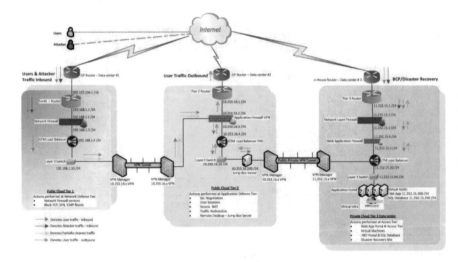

Fig. 7.37 Three-tier architecture design

This section further illustrates each tier in detail regarding implementation and setup (Fig. 7.38).

By running HPING command "*HPING3 -1 -C -K 3 --flood 202.122.134.1*" from a single system, potentially 180 Mbps bandwidth DoS attack is generated. By combining several Windows 7 systems as clients running the HPING, the output well be magnified. To detect and block HPING ICMP, network firewall (192.168.1.2) is reconfigured as:

```
Tier1NwFw(config-cmap)# icmp unreachable rate-limit 1 burst-size 1
Tier1NwFw(config-cmap)# icmp deny any time-exceeded
Tier1NwFw(config-cmap)# icmp deny any unreachable WAN
```

In order to view the traffic inspection map, the network firewall is reconfigured as:

```
Tier1NwFw(config-cmap)# show running-config class-map inspection_default
Tier1NwFw(config-cmap)# class-map inspection_default
Tier1NwFw(config-cmap)# match default-inspection-traffic
Tier1NwFw(config-cmap)# match access-list inspect
```

After network defense is completed, the traffic is route to the second tier using IPSec VPN configured as per the following configuration:

7.6.2.1 Application Defense Layer

Tier1-VPN(config-ikev1-policy)# encryption 3des SHA-1	** 3DES for encryption algo
Tier1-VPN(config-ikev1-policy)# hash md5	** MD5 for hash
Tier1-VPN(config-ikev1-policy)# authentication rsa-sig	** RSA for authentication
Tier1-VPN(config-ikev1-policy)# group 2	** Diffe–Hellman identifier
Tier1-VPN(config-ikev1-policy)# lifetime 86,400	** 24 h for SA lifetime
Tier1-VPN(config-ikev1-policy)# crypto ike1 enable Out-port	** Terminate VPN

The second tier is also implemented using public cloud and designed for providing application layer defense against the layer 7 attack like Slow POST, UDP Floods using Web Application Firewall (Imperva WAF), and F5 for advanced load balancing rules as illustrated in Fig. 7.39 below.

Devices and servers implemented at Tier 2 data center is illustrated below.

7.6.2.2 Access Layer

After the traffic is scanned for the DDoS attacks, the remaining traffic of authenticated, legitimate cloud service users is allowed to access the third tier for accessing the application by using the hardened access server. After dropping DDoS attack sessions and cleaning the traffic from network and application attackers, the remaining traffic is allowed to access the cloud application (11.252.15.251) (Figs. 7.40 and 7.41).

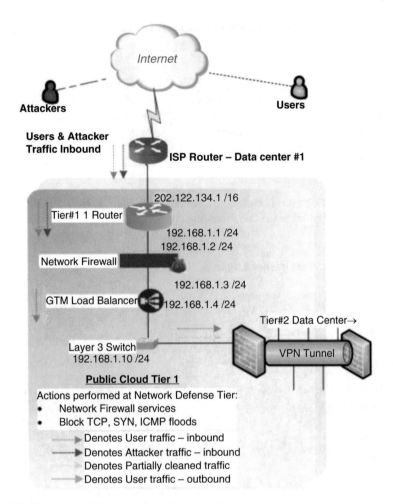

Fig. 7.38 Three-tier architecture—Network defense tier

7.7 Chapter Conclusion

7.7.1 Conclusion

The researcher introduced cloud Computing and DDoS, then research papers were reviewed and proposed a new DDoS attack classification, parameters for DDoS detection, and a new DDoS countermeasure taxonomy. Current cybersecurity threats and latest trends are illustrated, which also includes the DDoS-based survey performed by the researcher for gaining further insights. This lead to learnings that ransomware and DDoS are among the top concerns for organizations. Existing DDoS mitigation solutions ranging from On-Premise, ISP DDoS Service offerings to data scrubbing are reviewed and proposed a secure architecture design for miti-

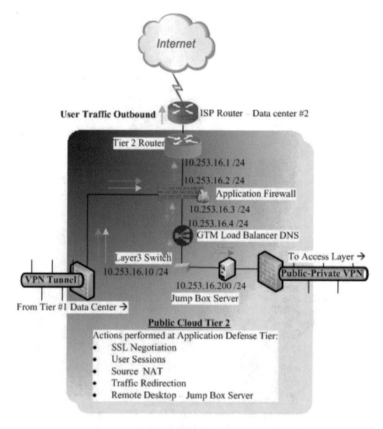

Fig. 7.39 Three-tier architecture—Application defense tier

gating DDoS attacks. The researcher implemented a secure infrastructure design in form of a three-tier architecture. DDoS attacks were performed on this infrastructure and compared with a single-tier standard architecture. The results obtained from the DDoS attacks are presented which clearly indicate that by providing multiple tiers of network and web application security in form of defense layers, it is possible to protect the availability, data integrity, and the performance of critical web applications, leading to higher customer confidence and lowering risks of underprovisioning the security and network devices. This is the main focus of this research chapter so that individuals and security administrators regarding the emerging threats and attacks using the proposed three-tier infrastructure design and validate results to arrive at accurate and knowledgeable judgement (Fig. 7.42).

Having only the network defense in place in form of firewall, the SaaS availability takes a major hit during DDoS attacks and the average real user response is always over 3700 ms—which indicates that the application availability is a huge issue. If using only the application defense, during DDoS attacks the average real user response goes even beyond 4100 ms—which clearly indicates that the application might not be available.

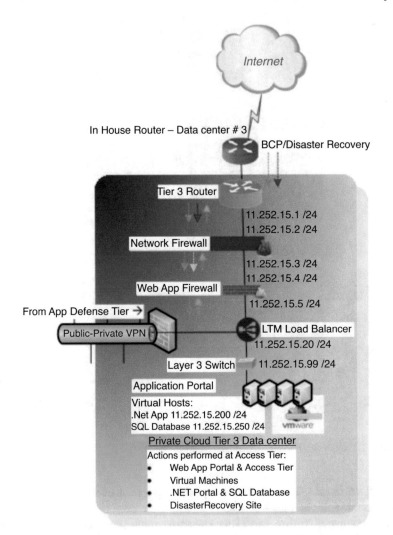

Fig. 7.40 Three-tier architecture—Access tier

As the size and scope of attacks increase the network devices like router and load balancer also start to show the strain, data for which is illustrated in Fig. 7.43 below, from CPU utilization %, process memory (5 min), and load balancer URL health.

Implementing the proposed three-tier architecture of network and application defense in form of separate tiers, for the same DDoS attack, the SaaS average real-time response is noticed to be around 1600 ms, which indicates that during the DDoS attack, the health remains in good shape and a resilient performance is guaranteed.

Attack#	Time (pm)	Buffer Size (bytes)	Echo Requests	Threads Count	Real User Monitoring				Status code	Attack Vector Details
					Average ICMP (ms)	Page Load Response (ms)	Browser Throughput (rpm)	App server response		
Attack#1	13:00	3700	1000	10	7655	50	1775	1528	200	No standard network or application layer defense in place three tier architecture Ping AppServer -n 1000 -l 3xxx Size: 3xxx, Echo request count: 1000
	13:30	3750	1000	15	7967	61	1826	1645	429	
	14:00	3760	1000	20	7202	70	1887	1517	200	
	14:30	3780	1000	25	7677	58	1773	1683	200	
	15:00	3790	1000	30	7993	65	1775	1692	429	
	15:30	3795	1000	35	6779	61	1850	1682	204	
	16:00	3800	1000	40	6016	63	1704	1534	429	
	16:30	3820	1000	45	7114	55	1804	1606	204	
	17:00	3810	1000	50	6242	50	1743	1547	503	
	17:30	3805	1000	55	7903	52	1751	1651	503	
	18:00	3820	1000	60	7766	72	1722	1685	503	
	18:30	3810	1000	65	6015	67	1860	1569	503	
	19:00	3805	1000	70	6042	64	1772	1674	503	
Attack#2	13:00	3700	1000	10	1746	11	1033	776	200	Network & Web ApplicationFirewall Defense implemented: Attack vector categories of attack as ICMP/UDP/SYN floods performed.
	13:30	3750	1000	15	1574	15	947	859	200	
	14:00	3760	1000	20	1548	11	935	850	200	
	14:30	3780	1000	25	1798	18	871	715	200	
	15:00	3790	1000	30	1795	18	1000	739	200	
	15:30	3795	1000	35	1549	15	888	736	200	
	16:00	3800	1000	40	1525	10	917	791	200	
	16:30	3820	1000	45	1827	12	878	807	200	
	17:00	3810	1000	50	1753	18	1029	768	200	
	17:30	3805	1000	55	1661	17	908	789	200	
	18:00	3820	1000	60	1733	11	1065	892	200	
	18:30	3810	1000	65	1685	17	1020	899	200	
	19:00	3805	1000	70	1536	11	1093	771	200	
	13:00	3700	1000	10	1697	16	906	701	200	
	13:30	3750	1000	15	1867	12	1028	823	200	
	14:00	3760	1000	20	1894	16	1016	857	200	
	14:30	3780	1000	25	1825	11	1093	710	200	

Fig. 7.41 Three-tier architecture attack logs

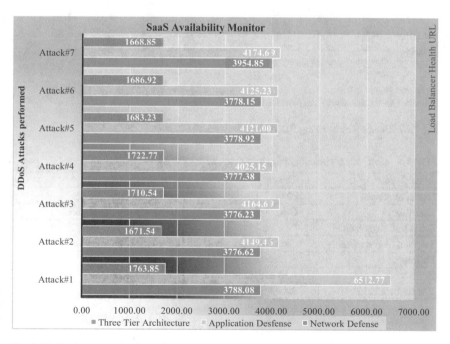

Fig. 7.42 Real user monitoring—SaaS availability threshold

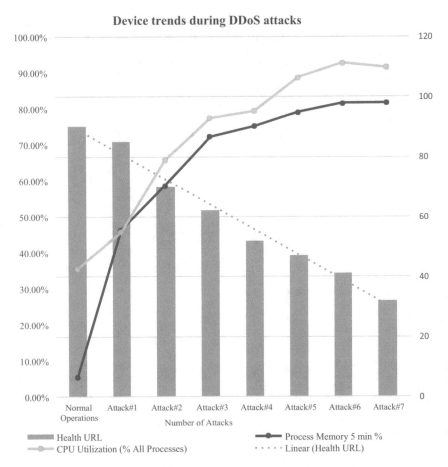

Fig. 7.43 Device trends during DDoS attacks

Table 7.6 Attacks in cloud computing infrastructure

Attack type	Category	Consequence
Denial of service	Cloud infrastructure	Service availability issue
Malware	Cloud infrastructure	Service availability issue
Cross VM side-channel	Cloud infrastructure	Information theft and leakage
Targeted shared memory	Cloud infrastructure	Cloud malware injection
Phishing	Access control	Unauthorized access
Botnets	Access control	Unauthorized access
VM rollback attack	Access control	Brute force launch, info leakage

7.7.2   Suggestions for Future Work

Cloud computing not only introduces additional risks and challenges but also adds various complications to deploying and maintaining the existing security standards. Widespread mobile device access and the on-demand services offered by cloud providers amplify the security concerns and threats even further. Table 7.6 above lists some of the known attacks and their consequences.

Given the various layers of cloud computing, security threats and DDoS attacks can be contained at different layers in the cloud computing environment. There are system-level threats, where an intruder bypasses the security to get unauthorized access, as well as cloud infrastructure- and network-level threats. Each component of a cloud should be separately addressed and requires equal attention to protect a cloud computing platform as a whole. The potential challenges in cloud computing can be categorized into the following four categories shown in Table 7.7 below.

These categories are closely related in various aspects and whenever one category is vulnerable to a certain attack, the other categories tend to fail ensuring the desired level of security. Thus, use of right set of security management design and level of precautions for a category can help strengthen the other categories a lot more and may eliminate the subsequent threats. As a result, security research in cloud computing should address the complete set of issues in a holistic approach, instead of an iterative or categorical resolution of threats and the distributed denial of service attacks.

Thus distributed denial of service attack mitigation has now become the topmost security priority for cloud server providers and cloud service consumers and there is a need for a systematic, verifiable and reliable cloud service delivery framework for cloud computing delivery to be sustainable. Given the complex operational structure of cloud computing frameworks, secure provenance of cloud-based data and services will always be a prominent research area in near future. Cloud computing has become the leading computational model for online application delivery. Due to the significant benefits in terms of flexibility, performance, and efficiency, cloud computing is slowly but steadily being adopted by almost all sectors. As more sectors migrate to cloud computing platforms, it becomes very important for cloud-based services to be fully ready for not only performance expectations but also for all types of potential security issues, risks, and challenges.

Table 7.7 Cloud computing categories and target areas

Category type	Areas targeted
Cloud infrastructure	Virtual machines, network and platform level
Access control	User- and resource-level authentication
Data outsourcing	Storage, transfer and data migration
Security standards	Cloud service agreements, SLA, implementations

References

A. Anwar, A. Sailor, K. Andrzej, O. Charles, A. Schulz, A. Segal Butt, Cost-aware cloud meter-
ing with scalable service management infrastructure, in *IEEE 8th International Conference
on Cloud Computing (ICCC)*, (2015), pp. 285–292. https://doi.org/10.1109/CLOUD.2015.46

S. Arukonda, S. Sinha, The innocent perpetrators: Reflectors skampskamp; reflection attacks. Adv.
Comput. Sci. **3**(13), 94–98 (2015)

M. Bhuyan, K. Bhattacharyya, K. Kalita, An empirical evaluation of information metrics for low
rate and high rate DDoS attack detection. Pattern Recognit. Lett. **51**, 1–7 (2015). https://doi.
org/10.1016/j.patrec.2014.07.019

H. Choi, S. Lim, B. Choi, R. Park, H. Lee, State of the art of network security perspectives in cloud
computing, in *Springer 1st International Conference on Communications in Computer and
Information Science Security-Enriched Urban Computing and Smart Grid*, (2010), pp. 629–
637. https://doi.org/10.1007/978-3-642-16444-6_79

M. Darwish, A. Ouda, F. Capretz, Cloud-based DDoS attacks and defenses, in *IEEE International
Conference on Information Society*, (2013), pp. 67–71. ISBN: 978-1-908320-13-1

R. Deshmukh, K. Devadkar, Understanding DDoS attack skampskamp; its effect in cloud environment, in *Procedia 4th International Conference on Advances in Computing, Communication and
Control (ICAC3'15)*, vol. 49, (2015), pp. 202–210. https://doi.org/10.1016/j.procs.2015.04.245

V. Durcekova, S. Ladislav, N. Shahmehri, Sophisticated denial of service attacks aimed at application layer, in *IEEE 9th International Conference 2012 ELEKTRO*, (2012). https://doi.
org/10.1109/ELEKTRO.2012.6225571

K. Georgios, M. Tassos, D. Geneiatakis, S. Gritzalis, A fair solution to DNS amplification attacks,
in *IEEE 2nd International Workshop on Digital Forensics and Incident Analysis (WDFIA)*,
(2007), pp. 38–47. https://doi.org/10.1109/WDFIA.2007.4299371

T. Hildmann, K. Odej, Deploying and extending on-premise cloud storage based on own cloud, in
IEEE 34th International Conference on Distributed Computing Systems Workshops (ICDCSW),
(2014), pp. 87–81. https://doi.org/10.1109/ICDCSW.2014.18

Imperva, 2016 Report. Imperva Global DDoS Threat Landscape Report (2016), https://www.
incapsula.com/DDoS-report/DDoS-report-q1-2016.html. Accessed 16 May 2016

B. Joshi, S. Vijayan, J. Kumar, Securing cloud computing environment against DDoS attacks,
in *IEEE International Conference on Computer Communication and Informatics*, (2012),
pp. 1–5. https://doi.org/10.1109/ICCCI.2012.6158817

A. Khadke, M. Madankar, M. Motghare, Review on mitigation of distributed denial of service
(DDoS) attacks in cloud computing, in *IEEE 10th International Conference on Intelligent
Systems and Control (ISCO)*, (2016), pp. 1–5. https://doi.org/10.1109/ISCO.2016.7726917

A. Malik, N. Muhammad, Security framework for cloud computing environment - a review.
J. Emerg. Trends Comput. Inform. Sci. **3**(3), 390–394 (2012) ISSN 2079-8407

A. Merlo, M. Migliardi, N. Gobbo, F. Palmieri, A. Castiglione, A denial of service attack to UMTS
networks using SIM-less devices. IEEE Trans. Depend. Sec. Comput. **11**(3), 280–291 (2014).
https://doi.org/10.1109/TDSC.2014.2315198

A. Mishra, A. Srivastava, B. Gupta, A. Tyagi, A. Sharma, A recent survey on DDoS attacks
and defense mechanisms, in *Springer 1st International Conference on Parallel, Distributed
Computing Technologies and Applications (PDCTA), (203)*, (2011), pp. 570–580. https://doi.
org/10.1007/978-3-642-24037-9_57

C. Prabhadevi, N. Syed, V. Sangeetha, Entropy-based anomaly detection system to prevent DDoS
attacks in cloud. Int. J. Comput. Appl. **62**(15), 42–47 (2014)

R. Sridaran, K. Nagaraju, The performance analysis of N-S architecture to mitigate DDoS attack in
cloud environment, in *IEEE 3rd International Conference on Computing for Sustainable Global
Development (INDIACom)*, (2016), pp. 3460–3463. ISBN: 978-9-3805-4421-2

F. Wong, X. Tan, A survey of trends in massive DDoS attacks and cloud-based mitigations. Int.
J. Netw. Secur. Appl. **6**(3), 57–71 (2014). https://doi.org/10.5121/ijnsa.2014.6305

T. Zargar, J. Joshi, T. David, A survey of defense mechanisms against distributed denial of service (DDoS) flooding attacks. IEEE Commun. Surv. Tutor. **15**(4), 2046–2069 (2013). https://doi.org/10.1109/SURV.2013.031413.00127

P. Zilberman, P. Rami, E. Yuval, On network footprint of traffic inspection and filtering at global scrubbing centers. IEEE Trans. Depend. Sec. Comput. **19**, 1–1 (2015). https://doi.org/10.1109/TDSC.2015.2494039

Chapter 8
Classifying Cyberattacks Amid Covid-19 Using Support Vector Machine

Jabeen Sultana and Abdul Khader Jilani

8.1 Introduction

With the advancement of technology, nowadays, cybersecurity is becoming very thought-provoking. It is regular for programmers, aggressors, and tricksters to capitalize on crises, especially in times when individuals are scared, edgy, and generally powerless during this pandemic. The flare-up of Covid isn't any unique. One of the underlying effects of COVID-19 is moving from the actual work environment to the web virtual work environment. This happened promptly over the world in numerous associations when the pandemic deteriorated and began influencing regular day-to-day existence. Coronavirus has pushed organizations to work rapidly and check framework strength as never done (Lallie et al. 2020). As the organizations are changing, new framework difficulties and needs are expanding, such as ongoing dynamic, online staff preparing, progression chances, and the biggest one is security risk. Agitators everywhere in the world are using the Covid as a fresh out of the plastic new device for their insidious deeds through hacking and attacking fake apps. Coronavirus pandemic has been found in an assortment of noxious missions including email misrepresentation, malware, ransomware, and malicious apps. As the number of cyber attacks increased in the current pandemic due to people browsing internet for Covid19 safety measures, causing burden by flooding thousands of computers and mobile devices with malicious atatcks (Hiscox 2019). These issues should be immediately fixed in the current circumstances (Jain et al. 2020).

It is common for hackers, attackers, and scammers to make the most of emergencies, particularly in times when people are frightened, desperate, and most

J. Sultana (✉) · A. K. Jilani
Department of Computer Science, College of Computer and Information Sciences, Majmaah University, Al Majmaah, Kingdom of Saudi Arabia
e-mail: J.sultana@mu.edu.sa; A.jilani@mu.edu.sa

© The Author(s), under exclusive license to Springer Nature Switzerland AG 2021
A. Bhardwaj, V. Sapra (eds.), *Security Incidents & Response Against Cyber Attacks*,
EAI/Springer Innovations in Communication and Computing,
https://doi.org/10.1007/978-3-030-69174-5_8

vulnerable during this pandemic. The outbreak of coronavirus is not any different. One of the initial impacts of COVID-19 is shifting from the physical workplace to the internet virtual workplace. This happened immediately across the world in many organizations as soon as the pandemic grew worse and started affecting everyday life. COVID-19 has pushed businesses to work quickly and check system durability as never done before (Lallie et al. 2020). As the businesses are transforming, new system challenges and priorities are increasing, for example real-time decision-making, online staff training, continuity risks, and the largest one is security risks. Bad actors all over the world are utilizing the coronavirus as a brand new tool for his or her evil deeds in the form of hacking, attacking, or frauds. COVID-19 pandemic has been found in a variety of malicious campaigns including email fraud, malware, ransomware, and malicious domains. As how many those afflicted continue to surge by thousands, campaigns that utilize the disease as a lure likewise increase. These issues need to be quickly addressed in the current situation (Jain et al. 2020).

The cause of increase in the crime rate was due to widespread of Covid-19. Internationally, "As of 3:33pm CET, 17 November 2020, there have been 54,771,888 confirmed cases of COVID-19, including 1,324,249 deaths, reported to WHO" (WHO 2020a). The organizations, which are in high risk of facing attacks in this COVID-19 pandemic, are Government Offices, Financial Services and Banking Systems and Health Care Systems. The various tools and devices used for this communication are ubiquitous in this current pandemic. Because of modern tools, people started communicating effectively throughout the globe. The online video conferencing applications such as Zoom, Microsoft Teams, and Google Meet have seen a remarkable expansion in new clients joining day by day. However, the usage of technology is bringing more issues and threats concerning cybersecurity (MalwareBytes 2020; The Times 2020), aiming to further strengthen the platforms resulting in hacking the data (Krebs on Security 2020; Smithers 2020), leading to scams by offering protection in person (Europol 2020). Many apps and websites were developed aiming to suggest tips and tricks to overcome this Covid-19, but in turn were actually intruding the systems, once the user downloads the app (Norton 2020; The Guardian 2020). Organizations must handle the growing security demands emerging from the increased risk of cyberattacks. They must also be mindful of the difficulties produced by the need to balance sensitive health information and privacy issues of men and women who might have been infected using them as they used because of heightened stress, anxiety, and worry facing individuals amid this pandemic (Nurse 2019).

Cyberattacks also targeted the software and tools in various sectors like health (Wired 2020). For instance, with the rapid growth of Zoom's popularity among people lead to wide counterattack as security professionals make alert that using Zoom is no longer safe as all the details are hacked. As a result, many companies such as NASA, SpaceX, and countries, including Taiwan, USA, and the Australian Defense force, banned Zoom for communication (Cross and Shinder 2008). Therefore, there is a need to realize these cyber threats and privacy concerns, which could result in unfavorable situations to mitigate or avoid them. A National Cyber Security Centre of U.K. bought to the public notice that how the criminals of cyber world are targeting the security systems of various nations in the current age of pandemic and ruining the secure information of almost all the sectors around the globe.

In this chapter, we aim to classify the tweets related to cybersecurity, causes, and attacks in order to bring awareness among the public because of this current Covid-19 pandemic.

8.2   Literature Review

With the broad adoption of digital technologies many facets of society have moved online, from shopping and social interactions to business, industry, and unfortunately, also crime. There were numerous cases where frauds and phishing tactics are circulating on Facebook Messenger and a number of other such applications because of increase in usage of internet technologies for meeting day-to-day activities like e-shopping, e-markets, and many more resulted in increase of crime rates. The frauds typically lure victims into free subscriptions such as Netflix premium free account. Once the victim clicks on the hyperlink, it redirects them with their social networking phishing website. In addition, the intruders attack the system using spam emails such as coronavirusfund@who.org. The WHO official website www. who.int ends with "int" and not with "org." Any user who did not confirm this email may turn into a victim. The cyberattacks occurring include DDOS attack, malicious websites and domains, malware, spam emails, and malicious messages on social networking, mobile threats and apps intended for browsing (Cross and Shinder 2008).

A digital attack was found to proliferate a fake COVID-19 data application that purportedly came from the World Health Organization. The programmer gets use of the switch Domain Name System setting in the D-Link or Linksys switches, which open the programs consequently and show a notification or a caution from the malicious application. The alarm just shows a choice named to download a "Coronavirus Inform application." When individual impulses on the download button, it introduces "Oski information stealer" malware on the gadget. This malware takes the programs' treats, put away passwords, program history and exchange data, and some more. It is hence amazingly trying for associations to create fitting assurance and reaction estimates given the changing situations as there is enormous expansion in cybercrime everywhere on the world. Late insights exhibit that there is fast expansion in cybercrimes and assessed to arrive at six trillion cases continuously till the year end. The reason for cybercrime includes around a casualty, a thought process, and a chance. The casualty is the objective of the assault, the rationale is the viewpoint driving the criminal to carry out the assault, and the open door is an opportunity for the wrongdoing to be perpetrated. Entrepreneurial aggressors consistently try to expand their benefit, and thusly, will trust that the best time will dispatch an attack where conditions fit those referenced previously. A disastrous event, progressing emergency or huge public occasion are ideal instances of these conditions (Tysiac 2018).

Bulk distractions interrupted globally, with individuals adjusting their day-by-day schedules to another reality: telecommuting, absence of social cooperation and actual action, and dread of not being readied (WHO 2020b; NHS 2020). These circumstances can overpower many, and cause pressure and tension that can expand the odds to be casualty of an attack. Likewise, the abrupt difference in working settings has implied that organizations have needed to extemporize new working structures, conceivably leaving corporate resources less ensured than before for interoperability. Since the COVID-19 began, the quantities of tricks and malware attacks have fundamentally risen (Gallagher and Brandt 2020), with phishing being accounted for to have expanded by 600% in March 2020 (Gallagher and Brandt 2020). During April 2020, Google apparently obstructed 18 million malware and phishing messages identified with the infection every day (Shi 2020). To improve probability of accomplishment, these attacks target offer of merchandise in particular which are popular in the name of Covid testing packs released by WHO and medications attracting people, but in turn ruining the public information with the help of these apps. In this way, intruders gain access to public details resulting in cyberattacks (Kumaran and Lugani 2020; O'Brien 2020).

The UK's Crown Prosecution Service (CPS) rules classify crimes in cyber world into two general classifications: "cyber-dependent and cyber-enabled crimes" (CPS 2019). Crimes that fall under cyber-independent category are like offense, "that must be submitted utilizing a PC, PC organizations, or other type of data correspondences innovation" (McGuire and Dowling 2013a). Crimes that fall under cyber-enabled category are "conventional wrongdoings, which can be expanded in their scale or reach by utilization of PCs, PC organizations, or different types of data correspondences innovation" (McGuire and Dowling 2013b). Pattern Micro Research as of late dissected a Covid themed malware that abrogates a frameworks' lord boot record, which makes it unbootable. The malware was definite in a public report distributed by the Czech network safety office. The malware document has "Covid Installer" in the portrayal. Once the malware executes, it will consequently restart the gear and afterward show an infection themed window that can't be shut. The run of the mill leave button on upper right half of the window doesn't work. This is the manner in which programmers gain section to the frameworks of obscure individuals and hack the subtleties.

In this paper, a technique was proposed based on "Auto-IF" for interruption identification dependent on profound learning approach utilizing auto encoder and isolation forest based on fog computing on the benchmark NSL-KDD dataset. The proposed technique accomplishes a high accuracy of 95.4% when contrasted with other techniques to identify the attacks (Sadaf and Sultana 2020). Aside from distinguishing and characterizing cyberattacks, AI procedures are likewise utilized in identifying other bad health condition types (Sultana and Jilani 2018; Sultana et al. 2019) and furthermore in grouping and considering informative opinions pertaining to education (Sultana et al. 2018).

In the current circumstances of deadly virus, the fraudsters are hacking emails of various organizations by using fake apps in the name of remedy for coronavirus and hacking the emails. The invaders first aim for the bank accounts. There are probabilities that they utilize the data of the clients and send them messages to improve their bank data and installment techniques as a result of the novel Covid. Applications declare that they include free facemask and security packs if application is downloaded. When an individual introduces the applying, this application conveys a SMSTrojan, which gathers the contact set of the casualty telephone catalog and sends auto SMS to spread itself. Crown veil offer introduces what has all the earmarks of being an innocuous malware which appropriates a SMS to all contacts. Apparently an update to the application will prepare the malware (Desai 2020). SMS requests that beneficiary take a required COVID-19 "planning" test, focuses to site which downloads malware and free school feast SMS guides beneficiary to site which takes installment certifications (Koenig 2020; Rodger 2020).

Basing the above literature, we thought of working on public sentiments of Covid-19 and how they fall into the trap of hackers by downloading malicious app unknowingly. Also, different kinds of attacks taking place in cyber world because of the current pandemic are analyzed and classified.

8.3 Proposed Method

The below five steps describe the suggested framework:

- Cybersecurity amid Covid-19 tweets are extracted using Twitter API.
- Tweets are well preprocessed using python libraries.
- Machine learning classifiers like SVM, Decision Tree, Hoeffding Tree, and Naive Bayes are used for training and testing the data with 70% of training data, a model is obtained.
- Test dataset with 30% of actual data is used for testing the obtained model and the results are obtained.
- Results obtained by the classifiers are evaluated using different metrics like accuracy, precision, recall, F-Score, ROC curve area, and K-statistics.

The below Fig. 8.1 describes the framework for classifying the cybercrime and attacks related tweets.

8.4 Methods

1. SVM: Support vector machines fall under the category of supervised learning techniques. SVMs work better for 2-class classification compared to 3-class classification. IT performs classification by constructing hyper plane and support vectors are constructed for the instances which fall under a particular class label. The manner in which SVM works is to plan vectors into an N-dimensional space and utilize an (N-1)-dimensional hyper plane as a choice plane to arrange information. The task of SVM is to locate the ideal hyper plane that isolates diverse

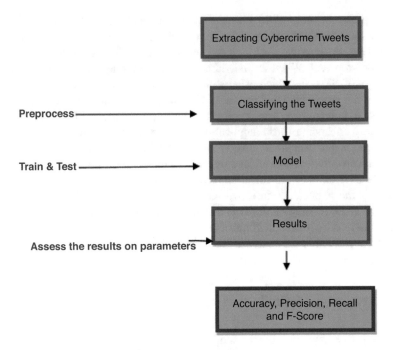

Fig. 8.1 A Framework to classify the cybercrime tweets

class enrollment. SVM goes through nonlinear planning to plan dataset into a high dimensional element space and utilize direct descriptor to characterize the information (Vapnik 1998).

2. Decision Trees: A basic classifier in machine learning. C4.5 utilizes the split-and-conquer method to create decision trees. These trees start at the base of the tree and go to its leaf center points. The J48 figuring that uses decision tree execution is used in the preliminaries itemized here. DT is commonly recognized in unique systems and is used because of its human sensible structure. A test thing for the class name starts from the base of the tree and goes through it to the leaf center, which gives the gathering of the model at its bottom node. A rule is calculated from top node or base node till bottom node (Quinlan 1993).

3. Hoeffding Tree: A Hoeffding tree is a gradually developed basing on decision tree that accomplishes to gain knowledge from information streams, expecting that the appropriation creating models don't change over the long run. Hoeffding trees achieve the way that a little example can regularly be sufficient to pick an ideal parting feature. This thought is represented in mathematical terms by the Hoeffding bound, which measures the quantity of perceptions expected to assess a few insights in classifying the data samples with high accuracy (Hulten et al. 2001).

4. Naive Bayes Tree: Naive Bayes Decision Tree (NBTREE) resembles DT beside at the leaves. It is a hybrid of decision tree and the Bayes rule is used to learn the

probabilities of each class using the given models. A given instance requires a measure of the unforeseen probability in order to be classified in a particular class label. Course of action at leaf center points is done by NB classifiers. Matched to DT, the NBTREE potential is accessible at each center point and the full scale probability isn't more than one (Kohavi 1996).

8.5   Results Discussion

In this section, we analyze the results. The below Table 8.1 describes the results of SVM, H-Tree, Naive Bayes, and Decision Tree.

8.5.1   Analyzing Results of Cyberattacks Amid Covid-19

The results are evaluated based on few parameters like accuracy, error rate, precision, recall, F-Score, Kappa statistic, and ROC curve. Results obtained by the machine learning classifiers like Decision Tree, H-Tree, and Naive Bayes are compared with SVM. Results are well analyzed and shown in the form of graphs in the below Figs. 8.2–8.7 and 8.8.

The below Fig. 8.2 describes the comparison of accuracies obtained among the classifiers.

SVM outperformed in classifying COVID tweets with highest accuracy of 94% followed by decision tree, H-Tree, and Naive Bayes with accuracies of 88%, 73%, and 51%. The below Fig. 8.3 describes the comparison of precision values obtained among the classifiers.

SVM obtained highest precision value with 0.94 followed by decision tree 0.88 and H-Tree attained 0.72. Naive Bayes with precision value of 0.50. The below Fig. 8.4 describes the comparison of recall values obtained among the classifiers.

Highest recall value was obtained for SVM 0.94 followed by decision trees, H-Tree, and Naive Bayes with recall values of 0.88, 0.73, and 0.51. The below Fig. 8.5 describes the comparison of F-Score values obtained among the classifiers.

Table 8.1   Results of Naive Bayes, Decision Tree, H-Tree, and SVM

Performance measures	Naive Bayes	Hoeffding Tree	Decision Tree	SVM
Accuracy	51	73	88	94
RMSE	0.44	0.32	0.24	0.20
Precision	0.50	0.72	0.88	0.94
Recall	0.51	0.73	0.88	0.94
F-Score	0.72	0.72	0.87	0.93
ROC	0.46	0.87	0.98	0.98
Kappa statistic	0.39	0.53	0.77	0.89

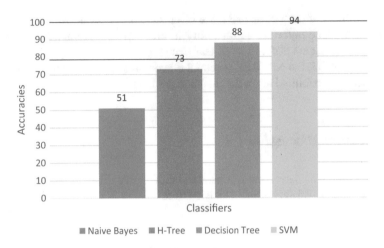

Fig. 8.2 Classification Accuracy on Cyberattack Tweets by different classifiers

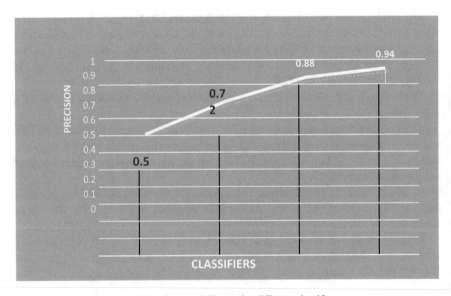

Fig. 8.3 Precision obtained on Cyberattack Tweets by different classifiers

Highest *F*-Score was obtained for SVM-0.93 followed by decision trees, H-Tree, and Naive Bayes and with recall values of 0.87, 0.72, and 0.72. The below Fig. 8.6 describes the comparison of root mean square error obtained among the classifiers.

Root mean square error was very low for SVM with 0.2 error rate followed by decision tree, H-Tree, and Naive Bayes and with 0.24, 0.32, and 0.44. The below Fig. 8.7 describes the comparison of ROC curve values and kappa statistic obtained among the classifiers (Fig. 8.8).

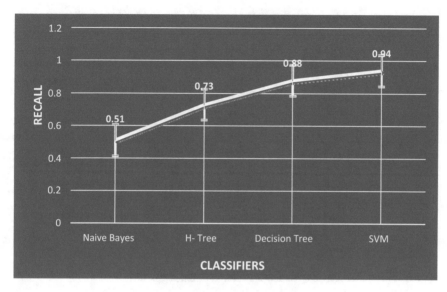

Fig. 8.4 Recall obtained on Cyberattack Tweets by different classifiers

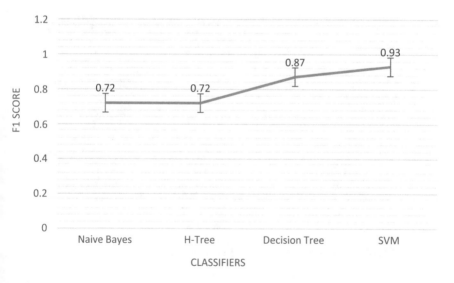

Fig. 8.5 *F*-Score obtained on Cyberattack Tweets by different classifiers

Also, threshold curve signifying ROC area is shown in the below Figs. 8.9–8.11 and 8.12 for the classifiers, namely SVM, decision tree, H-Tree, and Naive Bayes.

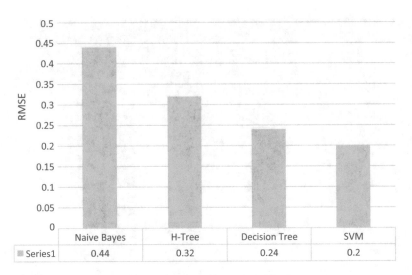

Fig. 8.6 RMSE obtained on Cyberattack Tweets by different classifiers

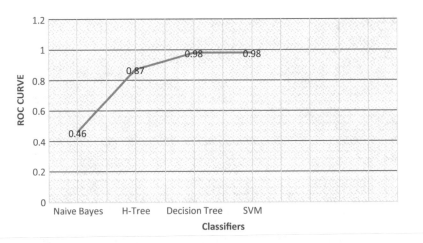

Fig. 8.7 ROC Curve values obtained on Cyberattack Tweets by different classifiers

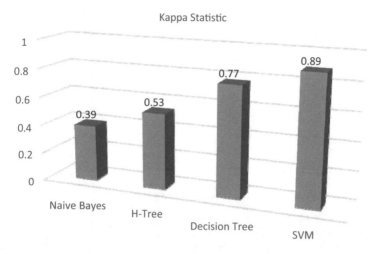

Fig. 8.8 Kappa Statistics obtained on Cyberattack Tweets by different classifiers

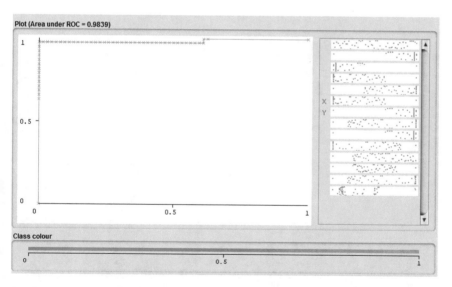

Fig. 8.9 Decision Tree threshold curve

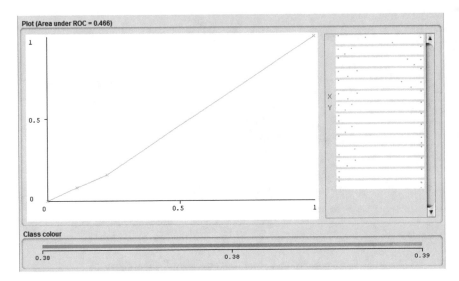

Fig. 8.10 Naive Bayes threshold curve

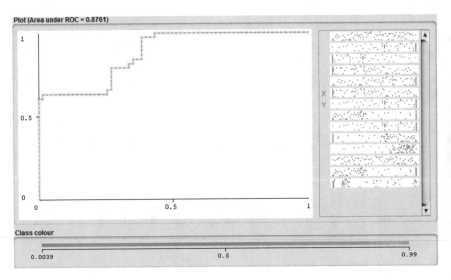

Fig. 8.11 H-Tree threshold curve

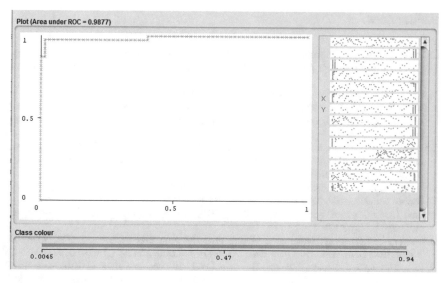

Fig. 8.12 SVM threshold curve

8.6   Conclusion

Due to the widespread of novel coronavirus, lockdowns emerged and social distancing was demanded and was the need to overcome the spread of this deadly virus. The more people started making use of network connections to meet their tasks, the more security systems have been vulnerable to cyberattacks. Anxiety, fear, and quench for finding out the cause and to overcome the spread of this Covid-19 made people to depend on more and more information worldwide. These days there are some websites named with corona aiming to give tips and tricks to overcome the current pandemic but in fact, installing them can infect the systems leading to system hang or crash. Therefore, it is concluded that intruders are targeting individuals, government officials, and even medical and health care systems in several ways and the organizations need to be alert to the security attacks by following preventive measures and installing strong antiviruses in the respective systems of the in-patient organizations.

The cybercrime related tweets of public in this Covid-19 age were extracted from twitter, preprocessed and classified well using machine learning classifiers. It was observed that support vector machine outperformed by yielding 94% accuracy in classifying the cybercrime related tweets with very less error rate. The results specify that the decision tree yields 88% classification accuracy followed by H-Tree and Naïve Bayes. Although, Naive Bayes is a good classifier but fails to produce good results here. Well-analyzed tweets reveal that the public tends to download malicious apps in the name of coronavirus tips and tricks in order to save lives and unknowingly fall in the trap of hackers leading to loss of crucial data. Therefore, public awareness has to be created among public to overcome cybercrimes. In future, we plan to improve the accuracy of Naive Bayes in classifying cybercrime tweets in this age of COVID-19.

References

CPS: "Cybercrime—prosecution guidance", The Crown Prosecution Service (CPS), Tech. Rep. (2019), https://www.cps.gov.uk/legal-guidance/cybercrimeprosecution-guidance. Accessed 17 Jun 2020

M. Cross, D.L. Shinder, *Scene of the Cybercrime* (Syngress, Boston, 2008)

S. Desai, New android app offers coronavirus safety mask but delivers SMS trojan (2020), https://www.zscaler.com/blogs/research/new-androidapp-offers-coronavirus-safety-mask-delivers-sms-trojan. Accessed 30 May 2020

Europol: Pandemic Profiteering: How Criminals Exploit COVID-19 Crisis (2020), https://www.europol.europa.eu/publicationsdocuments/pandemic-profiteering-how-criminalsexploit-covid-19-crisis. Accessed 15 Jun 2020

S. Gallagher, A. Brandt, Facing down the Myriad Threats Tied to covid19 (2020), https://news.sophos.com/enus/2020/04/14/covidmalware. Accessed 9 May 2020

Hiscox: The hiscox cyber readiness report 2019 (2019), https://www.hiscox.co.uk/cyberreadiness. Accessed 9 May 2020

G. Hulten, L. Spencer, P. Domingos, Mining time-changing data streams, in *ACM SIGKDD International Conference on Knowledge Discovery and Data Mining*, (2001), pp. 97–106

O. Jain, M. Gupta, S. Satam, S. Panda, Has the COVID-19 pandemic affected the susceptibility to cyberbullying in India? Computers in Human Behavior Reports **2**, 100029 (2020) ISSN 2451-9588

B. Koenig, Covid SMS Phishing Attempt (2020), https://twitter.com/BigBenKoenig/status/1242503232527589376. Accessed 30 May 2020

R. Kohavi, Scaling up the accuracy of Naïve-Bayes classifiers: A decision tree hybrid, in *Proceedings of KDD-96, Portland, USA*, (1996), pp. 202–207

Krebs on Security: Live Coronavirus Map Used to Spread Malware (2020), https://krebsonsecurity.com/2020/03/live-coronavirusmap-used-to-spread-malware/. Accessed 15 Jun 2020

N. Kumaran, S. Lugani, Protecting businesses against cyber threats during covid-19 and beyond (2020), https://cloud.google.com/blog/products/identitysecurity/protecting-against-cyber-threats-during-covid19-and-beyond. Accessed 17 Jun 2020

H.S. Lallie, L.A. Shepherd, J.R.C. Nurse, A. Erola, G. Epiphaniou, C. Maple, X. Bellekens, Cyber Security in the Age of COVID-19: A Timeline and Analysis of Cyber-Crime and Cyber-Attacks during the Pandemic, 2020

MalwareBytes: Cybercriminals impersonate World Health Organization to distribute fake coronavirus-book (2020), https://blog.malwarebytes.com/socialengineering/2020/03/cybercriminals-impersonate-worldhealth-organization-to-distribute-fake-coronavirus-ebook/. Accessed 15 Jun 2020

M. McGuire, S. Dowling, "Chapter 1: Cyberdependent crimes," Home Office, Tech. Rep. (2013a), https://assets.publishing.service.gov.uk/government/uploads/system/uploads/attachment_data/file/246751/horr75-chap1.pdf. Accessed 18 Jun 2020

M. McGuire, S. Dowling, "Chapter 2: Cyber-enabled crimes—fraud and theft," Home Office, Tech. Rep. (2013b), https://assets.publishing.service.gov.uk/government/uploads/system/uploads/attachment_data/file/248621/horr75-chap2.pdf. Accessed 18 Jun 2020

NHS: 10 tips to help if you are worried about coronavirus (2020), https://www.nhs.uk/oneyou/every-mindmatters/coronavirus-covid-19-anxiety-tips. Accessed 9 May 2020

Norton: Coronavirus phishing emails: How to protect against COVID-19 scams (2020), https://us.norton.com/internetsecurity-online-scamscoronavirus-phishing-scams.html. Accessed 15 Jun 2020

J.R.C. Nurse, Cybercrime and you: How criminals attack and the human factors that they seek to exploit, in *The Oxford Handbook of Cyberpsychology*, (OUP, Oxford, 2019)

T.L. O'Brien, Covid aid scams and dodgy 17 deals could have been avoided (2020), https://www.bloomberg.com/opinion/articles/2020-05-01/coronavirus-trillions-in-aid-draws-scams-anddodgy-deals. Accessed 9 May 2020

J.R. Quinlan, *C4.5 Programs for Machine Learning* (Morgan Kaufmann, Burlington, MA, 1993)

J. Rodger, The school meals coronavirus text scam which could trick parents out of thousands (2020), https://www.birminghammail.co.uk/news/midlandsnews/school-meals-coronavirus-text-scam-17975311. Accessed 30 May 2020

K. Sadaf, J. Sultana, Intrusion detection based on autoencoder and isolation forest in fog computing. IEEE Access **8**, 167059–167068 (2020). https://doi.org/10.1109/ACCESS.2020.3022855

F. Shi, Threat spotlight: Coronavirus-related phishing (2020), https://blog.barracuda.com/2020/03/26/threatspotlight-coronavirus-related-phishing. Accessed 9 May 2020

R. Smithers, Fraudsters use bogus NHS contact-tracing app in phishing scam (2020), https://www.theguardian.com/world/2020/may/13/fraudsters-use-bogus-nhs-contact-tracing-app-inphishing-scam. Accessed 30 May 2020

J. Sultana, A.K. Jilani, Predicting breast cancer using logistic regression and multi-class classifiers. Int. J. Eng. Technol. **7**(4.20), 22–26 (2018)

J. Sultana, N. Sultana, K. Yadav, F. Alfayez, Prediction of sentiment analysis on educational data based on deep learning approach, in *Proceedings of the 21st Saudi Computer Society National Computer Conference (NCC)*, vol. 1, (2018)

J. Sultana, K. Sadaf, A.K. Jilani, R. Alabdan, Diagnosing breast cancer using support vector machine and multi-classifiers, in *2019 International Conference on Computational Intelligence and Knowledge Economy (ICCIKE), Dubai, United Arab Emirates*, (2019), pp. 449–451. https://doi.org/10.1109/ICCIKE47802.2019.9004356

The Guardian: US Authorities battle surge in coronavirus scams, from phishing to fake treatments (2020), https://www.theguardian.com/world/2020/mar/19/coronavirus-scams-phishing-fake-treatments. Accessed 15 Jun 2020

The Times: Fraudsters impersonate airlines and Tesco in coronavirus scams (2020), https://www.thetimes.co.uk/article/fraudstersimpersonate-airlines-and-tesco-in-coronavirus-scams5wdwhxq7p. Accessed 15 Jun 2020

K. Tysiac, How cybercriminals prey on victims of natural disasters (2018), https://www.journalofaccountancy.com/news/2018/sep/cyber-criminals-prey-on-natural-disaster-victims201819720.html. Accessed 9 May 2020

V.N. Vapnik, *The Nature of Statistical Learning Theory*, 2nd edn. (Springer-Verlag, New York, 1998)

WHO: WHO Coronavirus disease (Covid-19) dashboard (2020a), https://covid19.who.int/. Accessed 15 Jun 2020

WHO: "#healthyathome" (2020b), https://www.who.int/news-room/campaigns/connectingtheworld-to-combat-coronavirus/healthyathome. Accessed 9 May 2020

Wired: Hackers are targeting hospitals crippled by coronavirus (2020), https://www.wired.co.uk/article/coronavirus-hackerscybercrime-phishing. Accessed 15 Jun 2020

Chapter 9
Cybersecurity Incident Response Against Advanced Persistent Threats (APTs)

Akashdeep Bhardwaj

9.1 Introduction

In today's era, internet usage has a vital role and a massive impact on all spheres of our professional work and personal lives. At the end user level, this has led to the need for having high level of privacy and security of our data, including personal information, documents, photographs, and videos which may be shared or kept private. Financial institutions like banks have started online banking commerce for daily operations, which again relies on internet. Government agencies and service organizations also rely on internet for their operations. This increasing involvement of the unsecure internet has cyber threats in the form of security issues and vulnerabilities. Unfortunately, most corporates and private organizations defend digital assets as a reactive process. Using traditional legacy security systems for analyzing the root cause of attacks has always been the standard norm. Instead of actually looking at the impact or proactively repairing the impact stemming from a security breach going for time-consuming, state of art remediation efforts to. The cycle persists because new systems and applications are constantly implemented across organization without actually planning early for cyber resiliency, which is a proactive approach at design stage of new systems or applications. This is as much about culture and education as it is about technology. Initially during the late 1990s to early 2000s, hackers were typically newbies displaying their newly acquired skills or simply causing mayhem for fun. From the year 2010, cyberattacks evolved into organized crime syndicates. Hackers started targeting vulnerable systems and users. This has now led to a highly profitable cybercrime industry thriving majorly on the dark web running illegal underground operations. Huge amount of state, corporate,

A. Bhardwaj (✉)
School of Computer Science, University of Petroleum and Energy Studies, Dehradun, India

© The Author(s), under exclusive license to Springer Nature Switzerland AG 2021
A. Bhardwaj, V. Sapra (eds.), *Security Incidents & Response Against Cyber Attacks*,
EAI/Springer Innovations in Communication and Computing,
https://doi.org/10.1007/978-3-030-69174-5_9

and personal data is constantly being accessed in unauthorized manner and hacked, only to be sold in these underground markets.

The strategies to harvest end user data or disrupt operations and services have evolved to a high level of sophistication with enhancements happening every day. This effectively indicates that the current attack methodology is most likely to be different in next year or even few months. Targeted cybersecurity attacks against corporate and public organizations, with time have increased and become highly extensive, sophisticated, and impactful. Even as security measures have enhanced, there is always a threat that warrants the security teams to be on full alert. At state and government level, such attacks range from Cyber Espionage, Cyber Extortion, Denial of Service, Zero Day Exploits, Identity Theft, Malware Ransomware, Man-in-the-Middle, and Phishing. Every year hackers devise advanced but potent attack methods. The most threating method for sustained access and infiltration is Advanced Persistent Threat or APT (Goldstein 2019), which is in fact a military term, adopted for cybersecurity context (Gilban 2019). This refers to attacks that are supported and performed by nation-states. Advanced persistent threats access unauthorized data and conceal themselves until the attack needs to be kept going. This involves a group of state-funded, highly skilled, determined and motivated advanced attackers. They use sophisticated methods and tools that use multiple attack vectors to ensure the attack bypasses and hidden from intrusion detection (KacyConnect 2019) and antivirus security systems. Most of the APTs follow low-and-slow approach that increases the success rate. This also includes identifying vulnerabilities of the security systems to create backdoors and exploits and gain entry into the organization and gather crucial and sensitive information. As per NIST, advanced level attacks (NIST Public Draft SP 800–160 Vol. 2 2019) then to hunt the targets repeatedly, which can be extended over a long duration. These attackers are well equipped to defend and find work around efforts to resist the sustained attacks and are resolute enough to preserve the same level probe and attack required to execute their malicious intents.

Organizations also adopt a patchwork defense posture with gaps that render them vulnerable against the next incursion. To become truly cyber resilient, a proactive approach will only take root if the following three steps are being followed:

- Board members and Higher Management should have a strong understanding of the cyber threat landscape faced by organization and stress on the importance of proactive cyber resilience culture.
- Cyber resilience policies and practices need to be implemented and driven by the Board leaders regularly via employee conversations to ensure stakeholders understand the consequences of cyber threats and their impact.
- New process, applications and system, by internal teams or external vendors, should establish cyber resilience from the project initiation until the final design and implementation.

However, proactive cyber resilience (NIST Cyber Resilience 2019) is impossible for any organization to achieve on its own. The dynamic and complex array of threats we face today demands each organization to seek a range of external perspectives on their cyber resiliency programs. Getting impartial and frank advice on your current ICT vulnerabilities and recovery plans is an essential component of any robust cyber resiliency program. However, in this global threat landscape, the most significant new constituent is the rise of state-funded international cyber sabotage operations. These are long-term, highly targeted campaigns known as Advanced Persistent Threats or APTs. Corporate organizations and public institutions have woken up to the malicious potential and impact of cybersecurity attacks. APTs are designed to evade organization's existing countermeasures like bypassing firewalls, IDS/IPS systems as well as anti-malware or threat hunting programs. APTs exploit vulnerabilities for launching their malicious incursions using 91% spear-phishing email attacks to infiltrate targets. Of these about 87% organizations fall for their malicious attacks. Unfortunately, there is no common understanding of what or how to respond to APTS and similar cybersecurity attacks. With no defined standards, organization end up adopting random practices and views and becomes increasingly important to plan effectively or to comprehend the incident response capability and the level of mitigation support required. While not every organization targeted by APTs, these attacks do present serious threat to specific organizations.

9.2   Advanced Persistent Threat Kill Chain

Traditional attacks target single person, mostly unspecified individuals for demonstrating capabilities or financial benefits, with the approach being to hit-and-run or smash-and-grab in a short period. APTs on the other hand are highly sophisticated and organized, operated by well-funded groups targeting high value state personalities, government institutions for strategic benefits and competitive advantages, involving repeated, determined attempts on long-term basis. The attackers employ multiple attack stages in various forms to stay hidden to perform the detrimental activities. The APT Kill Chain stages are illustrated in Fig. 9.1 below.

APTs are never a solitary stage attack, in fact it comprises of multiple attack vectors, hacking tools, and methodologies with several phases. The APT attackers possess huge resources at their disposal, are highly knowledgeable and skilled. Unlike traditional attacks, APTs are uniquely sophisticated. These attacks have clear goals and target to attack. These attacks constantly track the target over long periods of time, which varies few months few years. The attack is stealthy in approach and resilient enough to adapt against new security measures. APT Kill Chain attacks are performed in five stages as described below.

- **Stage#1: Infiltration or Reconnaissance**

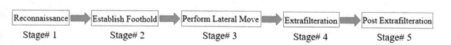

Fig. 9.1   Kill chain stages of advanced persistent threat attacks

This phase involves gathering information about the target. This step gathers as many vulnerabilities and exploits possible, applied for maximum impact. These involve basic footprinting techniques, Social Engineering and Open Source Intelligence (OSINT). There have been instances of attackers offering to buy sensitive information about the target, blackmail or coheres.

- **Stage #2: Establish Foothold or Infiltrate by breaking-in**
  This stage involves gaining access by exploiting the vulnerabilities of the target. This is done directly, which involves by working along inside the target organization, using insider access privilege to get access into the target systems using an infected USB while in indirect access, the attackers use emails for spear phishing. This involves emails having malicious script payloads and exploits sent as attachments, other methods include watering hole attacks to deploy Remote Access Trojans (RATs) to infiltrate further and performing high-level malicious activities.

- **Stage #3 Target Identification and Lateral Move**
  The third stage involves identifying lateral targets and move. During this stage, target scanning, abnormal traffic, or access violations are high-risk tasks and can result in being compromised.

- **Stage #4 Exfiltration**
  Once the target is scanned, the next step involves behavior obfuscation and privilege escalation using rootkits and access points to capture data following inside an organization's network. Such information captured gives the information to plan future attacks or staying hidden, which in turn ensure persistent access. Behavior obfuscation hides the behavior of malwares utilized in APT attacks.

- **Stage #5: Post Exfiltration**
  This is the final phase and involves sending the extracted data from the target to the command-and-control server, removing logs, covering tracks and escaping silently or staying hidden by maintaining backdoor access.

Advanced persistent threat tools (Partridge and Hendee 2018) have been known to be utilized initially during the infiltration stage or to access the target victim. The APT toolset can vary as usually the APT attackers develop and utilize their own tools and they have state resources to state of art technologies.

- LSB Steganography technique uses the science of hiding information into images, these are sent to unaware victims to initiate the initial APT phase as well as during the final phase to hide and extract the date back to the attacker group. APT groups use the least significant bit (LSB) in this toolset.
- Netbox RAT is actually a legal service which can gain access of the target system. This is also utilized by several organizations for providing IT support to remote workers and branch offices.
- LaLass is a logon session password cracking tool which can bypass hash attacks and then move horizontally inside the network without being detected or raising any alerts and red flags.
- HTRAN or HUC Packet Transfer Tool is a reverse proxy having the bouncer toolset. This has a listener application that silently gets inside the victim's system

anywhere over internet. Once the proxy listener starts to receive response from the actual target system, the responses are then redirected to the command-and-control server for the hacker to take over.

- Secure Delete is the final phase tool to securely delete logs and file in format followed by Department of Defense 522-22-M standard. This makes performing any log analysis, forensics, or data recovery difficult to perform.

9.3   Literature Survey

The author performed comprehensive survey and reviewed background of the literature within the topic area of advanced persistent threats for factual or nonfiction articles and publications. This proved to be the critical component of this chapter as part of research process. The survey helped understand the existing knowledge level and provided in-depth study of the recently published journal articles since 2019 to till date.

Advanced Persistent Threat (APT) assaults have caused genuine security dangers and money related misfortunes around the world. Different constant location components that join setting data and provenance diagrams have been proposed to guard against APT assaults. In any case, existing constant APT recognition instruments experience the ill effects of precision and productivity issues because of off base location models and the developing size of provenance charts. To address the exactness issue, Xiong et al. (2020) proposed a novel and precise APT identification model that expels pointless stages and spotlights on the staying ones with improved definitions. To address the proficiency issue, a state-based system is proposed in which occasions are devoured as streams and every element is spoken to in a FSA-like structure without putting away memorable information. Furthermore, the authors recreated assault situations by putting away only one of every a thousand occasions in a database. The system was tested on Windows and lead exhaustive tests under genuine situations to show that proposed research can precisely and productively distinguish all assaults inside our assessment. The memory utilization and CPU effectiveness remained consistent after some time (1–10 MB of memory and many occasions quicker than standard IT infra setups), making the proposed project a go-to solution for recognizing both known and unknown APT attacks in real-time scenarios.

Yuan et al. (2020) demonstrated the security issue for a cloud control framework. In the framework, advanced persistent threats (APTs) can be propelled by a malicious attacker to decrease the nature of administration of cloud and decay the framework execution further. To protect against APTs and make a security as an assistance plot, a safeguard needs to dispense guard asset to various units serving to various plants for improving the general framework execution when the cloud control obliges numerous physical plants at the same time. In the wake of watching the safeguard's activity, the assailant chooses which serve units to contain. Taking into account that both safeguard and assailant are dependent upon asset imperatives, the collaboration of different sides is demonstrated by a game. The ideal answers for two players under various kinds of spending limitations are examined. Recreation models and correlation results are given to confirm the fundamental consequences of the examination.

Yu et al. (2019) introduced a framework to recognize Advanced Persistent Threats (APTs) and remaking of assault situations. Because of the data asymmetry among assailants and safeguards, recognizing APT assaults stays to be a test. Rather than focusing on singular endeavors, associating the different phases of the assault is a generously increasingly attainable system. The creators first develop a rendition-based provenance chart by dissecting framework review logs. At that point, the creators utilized the standard-based procedure to locate a specific phase of and interface these assaults to produce an assault way by utilizing the connection between the data streams. Additionally, execution of the compaction procedure to pack the size of the chart for versatile scientific examination. An assessment of our methodology demonstrates that the proposed framework had the option to catch the key phases of APT battles with high accuracy and reproduce the subtleties of APT attacks.

Yang et al. (2019a) showed how the advanced persistent attacks for digital undercover work like espionage represented an incredible danger to current associations. So as to moderate the effect of APT on an association, all the undermined frameworks in the association must be isolated and recouped in an auspicious and viable manner. This article centers around the issue of modifying a powerful isolate and recuperation (QAR) conspire for an association with the goal that the APT effect is limited. In view of a novel hub level plague model describing the impact of the QAR conspire on the normal condition of the fundamental system, we gauge the normal effect of APT under a QAR plot. On this premise, we model the first issue as an ideal control issue. By utilization of ideal control hypothesis, we infer the optimality framework for the ideal control issue and in this manner present the idea of ordinary potential ideal (NPO) control. Next, through relative trials, we find that the NPO control beats a lot of heuristic controls. Consequently, the QAR conspire related with the NPO control is palatable as far as the viability of protecting against APT. At long last, we analyze the impact of certain elements on the normal APT effect under the NPO control. This article would be useful to safeguard against APT for digital secret activities.

Kim et al. (2019a, b) proposed Advanced Persistent Threat (APT) as one of the prime digital dangers that ceaselessly assault explicit targets exfiltrated data or obliterate the framework. Since the attackers utilize different instruments and strategies as indicated by the objective, it is hard to portray APT assault in a solitary example. Subsequently, APT assaults are hard to protect against with general countermeasures. In nowadays, frameworks comprise of different segments and related partners, which makes it hard to consider all the security concerns. Right now, propose a metaphysics information base and its structure procedure to suggest security necessities dependent on APT assault cases and framework area information. The proposed information base is isolated into three sections: APT metaphysics, general security information cosmology, and area explicit information philosophy. Every philosophy can assist with understanding the security worries in their insight. While coordinating three ontologies into the issue space philosophy, the fitting security necessities can be determined with the security prerequisites suggestion process. The proposed information base and procedure can assist with inferring the security necessities while thinking about both genuine assaults and frameworks.

Radhakrishnan et al. (2019) introduced their examination on late malware episodes and demonstrated that the current endpoint security arrangements are not powerful enough to verify the frameworks from getting traded off. The procedures, similar to code jumbling alongside at least one zero-days, are utilized by malware engineers for sidestepping the security frameworks. These malwares are utilized for enormous scale assaults including Advanced Persistent Threats (APT), Botnets, Crypto jacking, and so forth. Crypto jacking represents an extreme danger to different associations and people. We are outlining various techniques accessible for the identification of malware.

Advanced Persistent Threats (APTs) are exceptionally focused on advanced multi-arrange assaults, using multi-day or close to zero-day malware. Coordinated at internetworked PC clients in the working environment, their development and predominance can be ascribed to both socio (human) and specialized (framework shortcomings and deficient digital resistances) vulnerabilities. While numerous APT assaults fuse a mix of socio-specialized vulnerabilities, scholarly research and announced episodes to a great extent portray the client as the noticeable contributing element that can debilitate the layers of specialized security in an association. Nicho and McDermott (2019) investigated numerous elements of socio factors (non-specialized vulnerabilities) that add to the accomplishment of APT assaults in associations. Master interviews were directed with ranking directors, working in government and private associations in the United Arab Emirates (UAE) over a time of 4 years (2014 to 2017). In spite of basic conviction that socio factors get predominately from client conduct, our investigation uncovered two new components of socio vulnerabilities, in particular the job of authoritative administration, and natural elements which likewise add to the achievement of APT assaults. The creators indicated that the three measurements proposed right now help managers and IT work force in associations to actualize a proper blend of socio-specialized countermeasures for APT dangers.

In this research, Milajerdi et al. (2019) introduced HOLMES, a framework that actualizes another way to deal with the discovery of Advanced Persistent Threats (APTs). HOLMES is propelled by a few contextual analyses of genuine world APTs that feature some shared objectives of APT on-screen characters. Basically, HOLMES expects to deliver a recognition signal that demonstrates the nearness of a planned arrangement of exercises that are a piece of an APT battle. One of the primary difficulties tended to by our methodology includes building up a suite of strategies that make the location signal vigorous and dependable. At a significant level, the methods grew successfully influence the relationship between suspicious data streams that emerge during an aggressor battle. Notwithstanding its location ability, HOLMES is additionally ready to create an elevated level chart that outlines the aggressor's activities progressively. This diagram can be utilized by an expert for a successful digital reaction. An assessment of our methodology against some genuine world APTs shows that HOLMES can distinguish APT crusades with high accuracy and low bogus caution rate. The minimal significant level charts created by HOLMES successfully outline a progressing assault crusade and can help continuous digital reaction activities.

In order to research on the dangers in power network by utilizing heterogeneous information sources in power data framework, Liu et al. (2019) proposed APT assault discovery sandbox innovation and dynamic guard framework dependent on huge information investigation innovation. To start with, the document is reestablished from the mirror traffic and executed statically. At that point, sandbox execution was done to bring investigation tests into controllable virtual condition, and dynamic examination and activity tests were directed. Through examining the dynamic handling procedure of tests, different known and obscure vindictive code, APT assaults, high-hazard Trojan steeds, and other system security dangers were extensively distinguished. At long last, the risk evaluation of malevolent examples is helped out and pictured through the huge information stage. The outcomes show that the technique proposed right now successfully caution of obscure dangers, improve the security level of framework information and have a specific dynamic resistance capacity. Furthermore, it can viably improve the speed and precision of intensity data framework security circumstance forecast.

APT for information burglary represents a serious danger to distributed storage frameworks (CSSs). An APT entertainer may take important information from the objective CSS even in a key manner. To shield a CSS from APT, the cloud protector needs to powerfully apportion the restricted security assets to recuperate the undermined stockpiling servers, targeting moderating his complete misfortune. Li and Yang (2019) tended to this dynamic distributed storage recuperation (DCSR) issue by utilizing differential game hypothesis. Initially, by presenting a normal state advancement model catching the CSS's normal state development process under a blend of assault technique and recuperation procedure, the creators estimated the APT assailant's net advantage and the cloud safeguard's absolute shortfall. On this premise and in the most pessimistic scenario circumstance where the cloud protector accepts that the APT assailant has full information on his normal misfortune, at that point the creators diminished the DCSR issue to a differential game-theoretic issue (the DCSR issue) to describe the key cooperation between the two gatherings. Furthermore, essential condition for Nash harmony of the DCSR issue was inferred and in this manner present the idea of serious technique profile. Next, the author studied the basic properties of the serious technique profile, trailed by some numerical models. At that point, broad similar analyses were directed to show that the serious system profile is better than countless arbitrarily produced technique profiles in the feeling of Nash harmony arrangement idea. At long last, the creators quickly examine the practicability (versatility and attainability) of this paper. Our discoveries will be useful to improve the APT protection capacities of the cloud protector.

Digital security has gotten a matter of a worldwide intrigue, and a few assaults target modern organizations and administrative associations. The progressed relentless dangers (APTs) have risen as another and complex adaptation of multi-arrange assaults (MSAs), focusing on chosen organizations and associations. Current APT discovery frameworks center around raising the location alarms instead of foreseeing APTs. Gauging the APT stages not just uncovers the APT life cycle in its beginning periods yet additionally assists with understanding the assailant's systems and points. Ghafir et al. (2019) proposed a novel interruption recognition framework for APT identification and forecast. This framework experiences two primary stages; the first accomplishes the assault situation reproduction. This stage has a connection structure to interface the basic alarms that have a place with the equivalent APT battle. The connection depends on coordinating the properties of the basic alarms that are created over a configurable time window. The second period of the proposed framework is the assault unraveling. This stage uses the shrouded Markov model (HMM) to decide the no doubt arrangement of APT stages for a given grouping of associated cautions. Additionally, an expectation calculation is created to anticipate the following stage of the APT crusade in the wake of processing the likelihood of each APT stage to be the subsequent stage of the assailant. The proposed approach gauges the succession of APT stages with a forecast precision of in any event 91.80%. Furthermore, it predicts the subsequent stage of the APT battle with a precision of 66.50%, 92.70%, and 100% dependent on two, three, and four corresponded cautions, individually.

As APT assault is a unique digital weapon focused on the particular focuses in the internet. The complex assault strategies, long stay time, and explicit goals make the customary protection system incapable. In any case, most existing investigations neglect to think about the hypothetical displaying of the entire APT assault. Right now, Giban et al. (2019) set up a hypothetical system to describe a data put together APT assault with respect to the internal systems. Specifically, numerical structure incorporated the underlying section model for choosing the passage focuses and the focused on assault model for considering the knowledge gathering, technique basic leadership, weaponization and parallel development. Through a progression of reproductions, we locate the ideal applicant hubs in the underlying passage model, watch the dynamic difference in the focused on assault, demonstrate and check the qualities of the APT assault.

Fu et al. (2019) first broke down the attributes of Advanced Persistent Threat (APT). As indicated by APT assault, this paper built up a BP neural system streamlined by improved versatile hereditary calculation to foresee the security danger of hubs in the system. Also, determined the way of APT assaults with the most extreme conceivable assault. At long last, tests check the viability and rightness of the calculation by reenacting assaults. Trials show that this model can viably assess the security circumstance in the system, for the defenders to receive compelling measures safeguard against APT assaults, in this way improving the security of the system.

APTs having centered objective alongside cutting edge and tenacious assaulting abilities under extraordinary camouflage is another pattern followed for digital assaults. Danger insight helps in distinguishing and forestalling APT by gathering a large group of information and dissecting pernicious conduct through proficient information sharing and ensuring the security and nature of data trade. For better insurance, controlled access to knowledge data and a reviewing standard to reexamine the criteria in analysis for a security rupture is required. Chandel et al. (2019) examined a danger knowledge sharing network model and propose an improvement to build the productivity of sharing by reevaluating the size and synthesis of a sharing network. In view of different outside condition factors, it channels the low-quality shared insight by reviewing the trust level of a network part and the nature of a bit of knowledge. We trust that this examination can fill in some security holes to assist associations with settling on a superior choice in taking care of the ever-expanding and persistently evolving digital assaults.

Savvy implanted empower brilliant assembling is a significant foundation for future ventures. Expanding security dangers are upsetting the ordinary activities of keen assembling. As a novel sort of danger, a progressed steady risk (APT) has the novel highlights of solid disguise, inertness, and long haul ensnarement, which can infiltrate the center frameworks of savvy fabricating, particularly for shrewd installed frameworks, and cause extraordinary demolition from the digital side to physical side. Nonetheless, the current security plans can't give feasible asset the board, which causes the center framework in savvy fabricating not to perform manageable secure recognition and barrier against APTs. To address this test, Wu et al. (2019) proposed a supportable secure administration component for brilliant assembling against APTs. The proposed component incorporates two sections: supportable danger insight examination and practical secure asset. Feasible danger knowledge investigation gives economical revelation of the signs of potential APTs, which has highlights of a powerless sign, low connection, and moderate time variety. The maintainable secure asset gives profound and nonstop assurance to clever installed frameworks in savvy producing. The assessments show the protection capacities and the possibility of the proposed component.

It is assessed that by 2030, 1 of every 4 vehicles out and about will be driverless with appropriation rates expanding this figure generously throughout the following barely any decades. It is evaluated that by 2030, 1 of every 4 vehicles out and about will be driverless with selection rates expanding this figure considerably throughout the following not many decades. The advantages of driverless vehicles are all around reported; however, there are security dangers right now should be tended to. APTs are moden dangers across the board danger in the conventional LAN/WAN field and keeping in mind that no archived event of this risk in VANET right now exists, its solitary coherent that this will develop and influence VANET as the innovation gets conspicuous and assault vectors increment. An assault of this nature would have decimating ramifications for the clients of these vehicles as well as the overall population with the huge scale of attack, death toll becomes a real time probability. Abreu (2019) played out the exploration with goals to consider how AI procedures could help distinguish APT assaults inside VANET, check adequacy of danger location, and assault forecast, make a practical combined dataset comprising of both amiable and assault information, characterize and extricate information stream qualities that best mimic APT movement, recognize AI classifiers that best identify assaults inside the given dataset characteristics, order both assault and benevolent information utilizing AI classifiers advertisement at last dissect calculation test results dependent on distinguished key execution markers.

As of late, a sort of cyberattacks, known as cutting edge determined dangers, has brought about intense misfortunes to different associations, for example, governments and ventures. The APT has the attributes of long-term, complex assault implies, and solid capacity to disguise themselves, which make it hard to distinguish them. Because of the absence of legitimate intends to secure the Information-driven IoT (ICIoT), the ICIoT gadgets are incredibly powerless against APT assaults. Additionally, among the current APT discovery techniques, most specialists receive those that concentrate the highlights of various APT assaults, and the majority of the highlights extricated are neighborhood, which prompts the way that the related strategies have poor adaptability, in this manner lessening the exactness. Moreover, the attackers can without much of a stretch evade the discovery by changing the nearby highlights. Ma et al. (2019) examined that it is unavoidable that the contaminated host will create C&C correspondence with the direction and control server (C&C server), during the procedure of APT assaults, and the C&C area names are the scaffold interfacing the interior disease with the C&C server. Also, a specific APT assault of one assault family, which is the gathering of the equivalent APT assaults, will in general guide the C&C space names to a similar IP subnet. Under the supposition that the APT attackers have restricted assault assets, the connection between C&C space names of APT and IP subnet is inescapable for the APT assailants to get higher assault productivity, which prompts the compelling following of APT assault conduct. In this manner, the authors built an identification technique dependent on the area names' diagram structure. This discovery technique can improve the location proficiency in the data driven web, particularly for the IoT gadgets. What's more, simultaneously, the authors utilize a proper pruning procedure and a preprocessing strategy to lessen the size of information to be prepared and improve the computational effectiveness. This identification strategy can likewise lessen the discovery run.

Dangers that have been principally focusing on nation-states have extended the objectives to include corporate organizations and non-government divisions. Such dangers are notable as cutting edge tenacious dangers (APTs), are those that each country and entrenched association fears and needs to secure itself against. While country supported APT assaults will consistently be set apart by their advancement, APT assaults that have gotten conspicuous in corporate divisions don't make it any less trying for the associations. The rate at which the assault instruments and methods are developing is making any current safety efforts lacking. As safeguards endeavor to verify each endpoint and each connection inside their systems, assailants are finding better approaches to infiltrate into their objective frameworks. New sophisticated malware keep cropping up every day. These mark and conduct that is near ordinary, a solitary danger location framework would not get the job done. While it requires lots of persistence, patience, and time to perform APT, arrangements that adjust to the changing conduct of APT attacker(s) are required. A few works have been distributed on identifying an APT assault at a couple of its stages; however, exceptionally restricted research exists in distinguishing APT all in all from observation to cleanup, all things considered an answer requests complex connection and fine-grained conduct investigation of clients and frameworks inside and across systems. Through this overview paper, Alshamrani et al. (2019) reviewed every one of those strategies and procedures that could be utilized to identify various phases of APT assaults, learning techniques that should be applied and where to make your danger identification system savvy and obscure for those adjusting APT attackers. We additionally present diverse contextual analyses of APT assaults, distinctive observing techniques, and moderation strategies to be utilized for fine-grained control of security of an arranged framework. We close this paper with various difficulties in safeguarding against APT and open doors for additional examination.

Progressed relentless danger (APT) is another sort of cyberattack that represents a genuine risk to current society. At the point when an APT crusade on an association has been distinguished, the accessible fix assets must be sensibly apportioned to the conceivably unreliable hosts to moderate the potential loss of the association. Yang et al. (2019b) alluded to the achievable fix asset distribution methodologies as fix procedures. This exploration concentrated on the APT fix issue, i.e., the issue of creating viable fix techniques for associations. To begin with, for an association with time-differing correspondence relationship, we set up a development model of the association's normal state, where the effect of sidelong development of APT is obliged. On this premise, we model the APT fix issue as a differential Nash game issue (the APT fix game) in which the attacker endeavors to boost his latent capacity advantage, and the association figures out how to limit its potential misfortune. Second, we infer a framework (the potential framework) for ascertaining a potential Nash harmony of an APT fix game, and we analyze the structure of the potential assault and fix procedures in a potential Nash balance. Next, we unravel some potential frameworks to get the relating potential Nash equilibria. At long last, by examination with an enormous number of arbitrarily created assault and fix techniques, we reason that the potential Nash balance of each APT fix game is a Nash balance of the game. In this way, we prescribe to associations their particular potential fix systems. Our discoveries help to more readily comprehend and viably protect against APT.

Joined with a wide range of assault structures, progressed constant dangers (APTs) are turning into a significant risk to digital security. Existing security assurance works commonly either center around one-shot case, or separate location from reaction choices. Such practices lead to tractable examination, yet miss out key characteristic APTs determination and hazard heterogeneity. Li and Yang (2019) proposed a Lyapunov-based security-mindful safeguard component sponsored by risk insight, where strong guard technique making depends on obtained heterogeneity information. By investigating worldly development of hazard level, we propose need mindful virtual lines, which together with assault lines, empower security-mindful reaction among has. In particular, a long haul time normal benefit expansion issue is planned. We initially create hazard confirmation control strategy to suit hosts' hazard resistance and reaction limit. Under numerous attack assets, resistance control strategy is executed on two-arrange choices, including corresponding reasonable asset distribution and host-assault task. Specifically, dispersed sale based task calculation is intended to catch vulnerability in the quantity of settled assaults, where high level threats have attack datasets, which are organized.

9.4 APT Mitigation Taxonomy

Advanced persistent threat attacks mitigation requires defense-in-depth approach, across the entry and exit points at multiple levels. Event logs at each level and entry-exit points need to be co-related with threat hunting. This needs to be performed so that even if the APT does manage to bypass the initial entry level, the attack can still be detected or monitored at other levels and entry points. From initial footprinting, to monitoring and finally detection methodology with threat hunting. The authors propose APT mitigation taxonomy as footprinting, monitoring, detection, and threat intelligence. This is further illustrated in Fig. 9.2 below.

First classification to defend against APT attack is monitoring at multiple levels, entry and exit points of the network and systems. This involves monitoring each and every server, systems, and end user device for anomalies and behaviors regarding memory, disk, traffic, code execution, and logs, including endpoint security, firewall, and content filtering analysis.

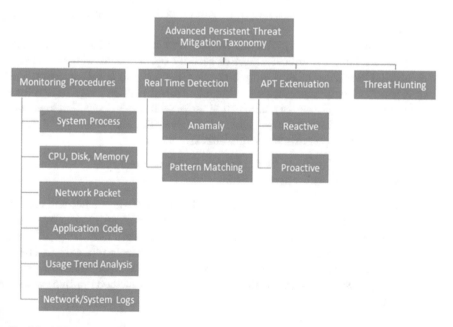

Fig. 9.2 APT Attack Mitigation Taxonomy

- Monitoring process includes ensuring OS and application patches are installed regularly as scheduled. Patches help remove the known vulnerabilities that may spread malware to other systems inside the network. Additionally, CPU usage monitoring also helps detecting high processing or spike which can evade detection by executing inside the system memory instead of executing from a file. These so-called fileless malwares use system process and vendor signed files found inside the OS and Apps that are already running within the memory itself. Since no additional separate process runs in the background, detection leaves no trace except for use of unexpected memory by the authorized process.
- Network traffic monitoring is critical for detecting APT attacks. These attacks use Command-and-Control Center Servers or C-n-C, which send instructions and payloads to compromised devices and systems multiple times or perform repeated data transmissions.
- Code Monitoring: Creating programming totally liberated from bugs is an illusion. Each product built up, each code that you discharge can never be destined to be sans blunder. While making the code itself mistake free is very troublesome, ensuring that it is sans blunder when running in various situations is unimaginable. These bugs are the methods for attackers to infiltrate frameworks. While some of them could be known before the code discharge, there is constantly a probability of obscure bugs. The potential vulnerabilities in the source code can be distinguished by static investigation procedures, for example, taint analysis and data flow examination. Likewise, checking the code during its execution for its exhibition and to ensure it runs inside its extension, neither using unforeseen assets nor spending memory areas that in any case are not available, would prompt recognizing a danger a lot prior before it can spread to different frameworks.
- Log Monitoring: Inbound traffic logs as well as event logs from servers, network devices are the most critical aspect of security operations as well as forensic analysis. Correlation of logs more often than not assist in attack detection and prevention at early analysis stage. When combined with end user device logs comprising of CPU, memory, application usage with system execution logs, this helps provide enormous information to analyze. The analysis helps defense against unknown system and network attacks.

Second classification from APT detection is by using real-time detection capabilities with anomaly and pattern matching.

- For anomaly detection the key feature to adapt for APTs is counter and defend against the attack. For this the attacker attempts and methods need to be monitored and learnt. This involves collection of logs and data from entry and exit sources, analyzing and making predictions from the data and then try to assess and respond to future attacks. However, such analysis methodologies are generally static checks. APT attackers tend to find evasion methods to bypass such detections.

- Conventionally cyberattack detections have relied majorly on signature and rule based pattern matching techniques. The problem with such approach is that only known attacks can be detected if the attack pattern is known and present in the security detection system. This approach is similar to virus detection—works only when the signature has been updated and present else the malware viruses can be bypassed.

Third classification involves extenuation by using reactive and proactive approach.

- Reactive approach categorizes the existing vulnerabilities and maps the IT Infrastructure and systems for calculating the attack surface scenarios. This assists in determining the possible attack scenarios and possible attack paths. Usually graph analysis is performed to map and analyze the data paths, complex networks, to identify sophisticated attack situations.
- Proactive approach is based on deception techniques. This tactic assists in tricking the attacker by manipulating the attack surface against evolving malwares or unknown and unexpected APT attacks. This helps to increase and preempt the difficulty level of attack by using Honeypot and Moving Targets in form of decoy portals, services, ports and even documents as well as having networks and systems simulating the production environments.

Fourth classification relates to use of threat hunting, this methodology is an iterative, focused tactic by defenders to examine, recognize, and comprehend the adversarial attack processes. These can be inside or outside the organization network. Since APT attackers are exceptionally skilled at obtaining undetected access, organization security teams often tend to remain unaware about such intrusions. This can go on for several days altogether or last for weeks and even months. Threat hunting starts with searching for Indicators of Compromise (or IOCs) and Indicators of Attack (or IOAs), which help identify unknown and existing attacks. Some of the common indicators are

- Unfamiliar inbound and outbound traffic at entry and exit points of the network
- Anomalies and irregular activities from admin and privileged level user accounts
- Impossible travel time due to geographical locations
- Risky login irregularities and consistent failures
- Database read query and volume increase
- Changes in HTML response magnitudes
- Increase in requests for same data files
- Mismatch in port-application request traffic
- Suspicious system configuration file and registry changes
- Anomalies in DNS request

9.5 Major APT Use Cases

This section presents few prominent APT attacks which were executed even before the term APT was reported or even coined.

- Titan Rain: In 2003, a progression of facilitated cyberattacks, later codenamed Titan Rain, have risen that penetrated a few PCs and systems related with U.S. Barrier Contractors with an objective to take touchy information. These were found to proceed until the finish of 2015, taking unclassified data from their objectives; however, no reports of taken ordered data were made. The degree of trickery included and the utilization of different assault vectors denoted these assaults as the first of their sort.
- Hydraq: One of the main APT assaults on business organizations that has drawn incredible consideration was Hydraq, name utilized in alluding to the Trojan that builds up the indirect access, notable under the first name given to this assault, "Activity Aurora." This organized assault included the utilization of a few malware segments that are scrambled in various layers to remain undetected for whatever length of time that they can. The assault saw as propelled in 2009 has focused on various association segments, Google being one of them and the first to report it, trailed by Adobe. The name "Aurora" originated from the references in the malware that got infused during the malware's gathering on the assailant's machine. The malware was found to utilize a zero-day misuse in Internet Explorer (CVE-2010-0249 and MS10-002) to build up a dependable balance on the framework. At the point when clients visited the pernicious webpage, Internet Explorer was misused to download a few malware segments. One of the malware parts set up a secondary passage to the machine, permitting attackers to get onto the association's system as and when required. In a portion of the previous cases, the malware misused a weakness in Adobe peruser and aerialist applications (CVE-2009-1862) to build up a dependable balance on not many organizations. In contrast to the prior occurrences, the later cases of these malware were found to never again utilize the zero-day vulnerabilities. By and by, the assaults proceeded for a while after, in various nations over the globe under various variations of the Trojan Hydraq. The basic part of the Trojan is, the malware assembles framework and system data at first, trailed by gathering usernames and secret key into a record that is later sent to its order and control focus whose IP address or area name is hard-coded inside the malware.
- Stuxnet: In 2009, a refined worm that spreads itself to different segments in the element with an objective to obstruct Iran's uranium atomic venture, had been propelled. From the start, this malware was found to misuse a zero-day powerlessness found in LNK document of Windows pilgrim. Microsoft named this malware as Stuxnet from a blend of record names found in the malevolent code (.stub and MrxNet.sys) subsequent to being accounted for about this zero-day helplessness. In any case, it was later discovered that notwithstanding the LNK

weakness, a helplessness in printer spooler of Windows PCs was utilized to spread across machines that common a printer. And afterward this malware utilized 2 vulnerabilities in Windows console record and Task Scheduler document to deal with the machine by performing benefit acceleration. Moreover, it utilized a hard-coded secret word inside a Siemens Step7 programming to contaminate database servers with Step7 and from that point taint different machines associated with it. After the malware first enters a framework, it sends the inner IP and the open IP of that framework alongside the PC name, working arrangement of the framework, and whether Siemens Step7 programming was introduced on that machine, to one of its 2 direction and control focuses running in 2 distinct nations. It was before long discovered that Stuxnet was path out of hand with a few PCs in various nations being tainted with this malware. After security specialists over the globe have delved into Stuxnet for a while, it was discovered that this malware was path past what it resembled and it really sends directions to programmable legitimate controllers focused to obstruct the Iran's uranium atomic undertaking. A few reports were distributed by scientists and firms over the world, with pretty much clashing data on the definite execution of Stuxnet. In any case, they all concur that Stuxnet was seen as more than ever, a ruin that a computerized code could make in physical world. It was not only around 4 zero-day vulnerabilities, 2 taken testaments, and 2 direction and control focuses, it was more than that, a cunningly created, layered bit of malware that could be changed by the attackers through the order and control focuses utilizing more than 400 things in its design record. The end date of Stuxnet was seen as in 2012, 3 years after it was released. In spite of the fact that Iran discovered the presence of this 500 KB malware in its Natanz plant in 2010, in the midst of all the devastation of Stuxnet, a portion of its axes were at that point harmed, hindering its atomic weapon age process.

- RSA SecureID Attack: In 2011, RSA, a protected division of EMC Software declared an advanced digital assault on its frameworks that included the tradeoff of data related with its SecureID, a 2 factor validation token item. This is another assault that penetrated an association's system through phishing messages sent to the association's bosses. As a component of this assault, the attackers sent 2 distinctive phishing messages to various gatherings of businesses with an exceed expectations sheet appended. The phishing messages went into the garbage organizer on the businesses end, in any case, they were created all around ok that a worker opened the connected exceed expectations sheet. This exceed expectations sheet when opened adventures the zero-day weakness (CVE-2011-0609) of adobe streak player to introduce an indirect access. At the point when the worker opened the previously mentioned connection, the secondary passage got introduced onto the representative's framework. This introduced secondary passage was seen as a variation of a notable remote organization instrument that now the assailants could use to remote access the representative's machine. With this remote access setup, the attackers began collecting accreditations of a few representatives with an end goal to arrive at the objective framework where they performed benefit accelerations, took the information and records, packed and encoded them before sending them to their remote order and control focus by means of ftp. RSA distinguished this exfiltration however not before a portion of the information got exfiltrated.

- Carbanak: Carbanak, not at all like the APT assaults examined before, was an assault for taking cash from money-related foundations. The assaults began in 2013, with the assailants getting into the inside system of their objective banking/money related establishment through lance phishing assaults, have gone undetected until mid-2014. As indicated by [30], messages sent to workers had records joined to them that when executed adventure Microsoft office vulnerabilities (CVE-2012-0158, CVE-2013-3906, and CVE-2014-1761) giving the pernicious code in the connection capacity to introduce indirect access. This indirect access was named Carbanak apparently after, Carberp, the malware utilized as secondary passage that was viewed as a variation of a known malware "Anunak." When the assailants set up an a dependable balance, they began inside surveillance as a feature of their parallel development, through key lumberjacks, structure grabbers, and such which were sent to the attackers C&C server. Scientists discovered recordings of representatives exercises were caught and sent to the assailant's direction and control server as a component of their interior surveillance. The concern now is for various instruments they utilized, and a custom twofold convention they set up to speak with their C&C servers from the exploited people machines. It was discovered that the attackers examined every one of their unfortunate casualties through their inner observation and utilized assault strategies that would explicitly apply to that injured individual. They made phony exchanges in the injured individual's inside database to conceal their cash move exchanges. Carbanak appeared to stop in 2015, just to be discovered later that it kept on appearing through 2017 in changed variations.

The above APT assaults tend to be seen that as the years passed by with an ever increasing number of advanced components turning out to be a piece of the physical world, the degree of trickiness included and the strategies utilized have expanded in their complexity. Symantec, in its 2018 Internet Security Threat Report, Volume 23, examined developing types of digital wrongdoing that will in general be focused on assaults in the veil of ransomware. As indicated by this report, directed assault bunches are utilizing ransomware as fake to perform unfavorable exercises on their objectives. NotPetya, one of such assaults, was accounted for by ESET, an IT security organization, as crafted by one of the advancing APT gatherings whom they call Telebots. In light of present conditions, so as to guard against APT assaults a safeguard top to bottom methodology that figures out how to adjust to the attackers' strategies should be created, which we talk about in the following area. The contextual analyses point not exclusively to the requirement for smart barrier systems, yet additionally to the extent of these APT assaults. In spite of the fact that these APTs have begun in country state areas, it didn't set aside a lot of effort for the assailants to stretch out their extension to nongovernmental and business segments with objectives of taking corporate information representing the greatest risk to any organization with information as their greatest resource. Nonetheless, the later progressed determined dangers bring up that associations with resources other than information, for example, account associations where cash is the significant resource, are likewise confronting these dangers. The Carbanak assault talked about for our situation study is one such model.

9.6 Conclusion

Advanced persistent threats are looked upon and rated with the highest security concern. APTs comprise of determined, persistent and well-funded groups with target of accessing crucial data from high value individuals at critical state level organizations and government levels. APT toolsets and classification are discussed in this chapter, comparing with the targeted attacks. Use of advanced cybersecurity methodologies may be able to monitor and at times even identify the initial reconnaissance, network scans, and subsequent data loss. However, detecting the initial infiltration, lateral move inside the network, establishing foothold to staying hidden is difficult to detect. As most end users and organizations are starting to adopt IoT devices, more advanced level of measures and defenses are required against APT attacks.

References

S. Abreu, A feasibility study on machine learning techniques for APT detection and protection in VANETs, in *IEEE 12th International Conference on Global Security, Safety and Sustainability (ICGS3), London, United Kingdom*, (2019). https://doi.org/10.1109/ICGS3.2019.8688031

A. Alshamrani, S. Myneni, A. Chowdhary, W. Huang, Survey on advanced persistent threats: Techniques, solutions, challenges, and research opportunities. IEEE Commun. Surv. Tutor. **21**(2(Second Quarter 2019)), 1851–1877 (2019). https://doi.org/10.1109/COMST.2019.2891891

S. Chandel, M. Yan, S. Chen, H. Jiang, T. Ni, Threat intelligence sharing community: A countermeasure against advanced persistent threat, in *IEEE Conference on Multimedia Information Processing and Retrieval (MIPR), San Jose, CA, USA*, (2019)

T. Fu, Y. Lu, W. Zhen, APT attack situation assessment model based on optimized BP neural network, in *IEEE 3rd Information Technology, Networking, Electronic and Automation Control Conference (ITNEC), Chengdu, China*, (2019). https://doi.org/10.1109/ITNEC.2019.8729178

I. Ghafir, G. Kyriakopoulos, S. Lambotharan, J. Francisco, Hidden Markov models and alert correlations for the prediction of advanced persistent threats. IEEE Access **7**, 99508–99520 (2019). https://doi.org/10.1109/ACCESS.2019.2930200

M. Gilban, How advanced persistent threats work (2019), https://xmcyber.com/how-advanced-persistent-threats-work/. Accessed 7 Feb 2020

J. Goldstein, What are Advanced Persistent Threats (APTs), and how do you find them? (2019), https://securityintelligence.com/posts/what-are-advanced-persistent-threats-apts-and-how-do-you-find-them. Accessed 5 Feb 2020

K.Z.N.W.E. KacyConnect, APT uses arsenal of tools to evade detection (2019), https://www.infosecurity-magazine.com/news/apt-uses-arsenal-of-tools-to-evade-1/. Accessed 10 Jan 2020

M. Kim, S. Dey, S. Lee, Ontology-driven security requirements recommendation for APT attack, in *2019 IEEE 27th international requirements engineering conference workshops (REW), Jeju Island, South Korea*, (2019a). https://doi.org/10.1109/REW.2019.00032

Y. Kim, W. Dai, J. Bai, X. Gan, J. Wang, X. Wang, An intelligence-driven security-aware defense mechanism for advanced persistent threats. IEEE Trans. Inform. Forens. Secur. **14**(3), 646–661 (2019b). https://doi.org/10.1109/TIFS.2018.2847671

P. Li, X. Yang, On dynamic recovery of cloud storage system under advanced persistent threats. IEEE Access **7**, 103556–103569 (2019). https://doi.org/10.1109/ACCESS.2019.2932020

D. Liu, H. Zhang, H. Yu, X. Liu, X. Zhao, G. Lv, Research and application of APT attack defense and detection technology based on big data technology, in *IEEE 9th International Conference on Electronics Information and Emergency Communication (ICEIEC), Beijing, China*, (2019). https://doi.org/10.1109/ICEIEC.2019.8784483

Z. Ma, Q. Li, X. Meng, Discovering suspicious APT families through a large scale domain graph in information-centric IoT. IEEE Access **7**, 13917–13926 (2019). https://doi.org/10.1109/ACCESS.2019.2894509

S. Milajerdi, R. Gjomemo, B. Eshete, R. Sekar, N. Venkatakrishnan, HOLMES: Real-time APT detection through correlation of suspicious information flows, in *IEEE Symposium on Security and Privacy (SP), San Francisco, CA, USA*, (2019). https://doi.org/10.1109/SP.2019.00026

M. Nicho, C. McDermott, Dimensions of 'socio' vulnerabilities of advanced persistent threats, in *2019 IEEE International Conference on Software, Telecommunications and Computer Networks (SoftCOM), Split, Croatia*, (2019). https://doi.org/10.23919/SOFTCOM.2019.8903788

NIST Publishing APT Cyber Resilience Guidance in September (2019), https://www.meritalk.com/articles/nist-publishing-apt-cyber-resilience-guidance-in-september/. Accessed 21 Dec 2019

NIST Releases Final Public Draft SP 800-160 Vol. 2 (2019), https://csrc.nist.gov/News/2019/nist-releases-final-public-draft-sp-800-160-vol-2. Accessed 4 Nov 2019

C. Partridge, N. Hendee, From bear to vault: Designing a new protocol to extend the APT communications toolset, in *IEEE International Conference on Computational Science and Computational Intelligence (CSCI), Las Vegas, USA*, (2018). https://doi.org/10.1109/CSCI46756.2018.00028

K. Radhakrishnan, R. Menon, H. Nath, A survey of zero-day malware attacks and its detection methodology, in *IEEE TENCON Conference, Region 10, Kochi, India*, (2019). https://doi.org/10.1109/TENCON.2019.8929620

J. Wu, M. Dong, K. Ota, J. Li, W. Yang, Sustainable secure management against APT attacks for intelligent embedded-enabled smart manufacturing. IEEE Trans. Sustain. Comput. **5**, 341–352 (2019). https://doi.org/10.1109/TSUSC.2019.2913317

C. Xiong, T. Zhu, W. Dong, L. Ruan, R. Yang, Y. Chen, Y. Cheng, CONAN: A practical real-time APT detection system with high accuracy and efficiency, in *IEEE Transactions on Dependable and Secure Computing (Early Access)*, (2020). https://doi.org/10.1109/TDSC.2020.2971484

L. Yang, P. Li, X. Yang, Y. Xiang, F. Jiang, W. Zhou, Effective quarantine and recovery scheme against advanced persistent threat, in *IEEE Transactions on Systems, Man, and Cybernetics: Systems, Early Access*, (2019a). https://doi.org/10.1109/TSMC.2019.2956860

L. Yang, P. Li, Y. Zhang, X. Yang, Y. Xiang, W. Zhou, Effective repair strategy against advanced persistent threat: A differential game approach. IEEE Trans. Inform. Forens. Secur. **14**(7), 1713–1728 (2019b). https://doi.org/10.1109/TIFS.2018.2885251

H. Yu, A. Li, R. Jiang, Needle in a haystack: Attack detection from large-scale system audit, in *IEEE 19th International Conference on Communication Technology (ICCT), Xi'an, China*, (2019). https://doi.org/10.1109/ICCT46805.2019.8947201

H. Yuan, Y. Xia, J. Zhang, H. Yang, M. Mahmoud, Stackelberg: Game-based defense analysis against advanced persistent threats on cloud control system. IEEE Trans. Ind. Inform. **16**, 1571–1580 (2020). https://doi.org/10.1109/TII.2019.2925035

Chapter 10
IoT Architecture Vulnerabilities and Security Measures

Gaytri Bakshi

10.1 Introduction

IoT stands for the Internet of Things that includes micro to macro devices/things which are connected to each other via the internet, which is eventually a network of networks, making the whole planet as a home. IoT has a multi-layered architecture, which supports low energy consuming smart devices and various protocols that differs from the web stack. Focusing on the main objectives of smart, intelligent, innovative and power-efficient smart infrastructure; smart devices are proposed and built in order to reduce the human intervention and make the machine intelligent enough to perform the assigned tasks and help the humankind. Embedded smart sensors interact with the physical world to generate data in a readable form for the digital machines to apply analytics on it and invoke the decision-making process to generate suitable solutions for humans or actuate other digital equipment as per the requirements of the application. This whole process is supported by various protocols related to connectivity, data transfer, communication, device management, etc. but all of these are vulnerable to malevolent attacks. The focus of the chapter is to provide details about the upcoming or prevalent threats or issues related to the IoT security related to IoT devices security, data security and networks security and measures to counter them.

G. Bakshi (✉)
School of Computer Science, University of Petroleum and Energy Studies, bidholi, Dehradun, India

© The Author(s), under exclusive license to Springer Nature Switzerland AG 2021
A. Bhardwaj, V. Sapra (eds.), *Security Incidents & Response Against Cyber Attacks*, EAI/Springer Innovations in Communication and Computing, https://doi.org/10.1007/978-3-030-69174-5_10

10.2 Analysis of Literature

Science with its ventures have given human the support to develop a platform where every task is accomplished by machines and this was not possible in 1 day. It started in the late eighteenth century when communication over a distance was accomplished by the invention of the telegraph. Gradually, the origination of various electronic gadgets over the period of time involved the integration of sensors, computational machinery, and intelligence with the connected network used as the backbone for communication which was later termed as the Internet of Things by Kevin Ashton in 1999 ("A Brief History of the IoT" 14 May 2015). Internet as the major platform of interconnected network nodes constitutes several users in various arenas combining them in one system (Nunberg 2012). Progressively Internet of Things has endured to be cutting-edge, technology in not just the IT world but even in many interdisciplinary domains. It has fascinated and given a global platform to bind physical objects with one another and had led to the communication of humans to things and things to things and vice versa (Kosmatos et al. 2011). IoT framework actually contains sensor nodes, that may be remotely deployed and communicates over a wireless network with different analytics applied to data acquired by them and depending upon the application the results are provided to the user using third-party services and sometimes the actuator is triggered to perform specific action depending upon the application (Akyildiz et al. 2002). This scenario can be depicted in Fig. 10.1 below.

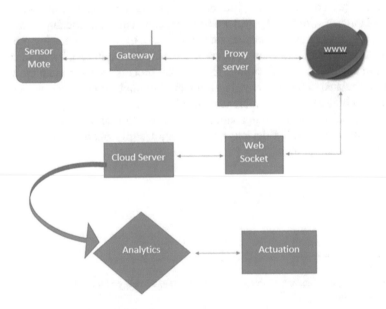

Fig. 10.1 IoT framework implementation

The entire framework does not work as a standalone system; there are enabling technologies that support the entire system to work as one. These technologies include the following:

(a) Big data: Big data refers to the collection of huge data sets that are collected by the sensor nodes and the aggregated data is stored, interrogated, examined, and managed in a way where analytics is performed on to it and it holds the metric such as:

- Volume: specifies the amount of data collection through IoT sensory nodes.
- Velocity: Specifies the speed and duration according to which the data is aggregated.
- Variety: Specifies the type of data that is collected and is segregated based on its types.

   Cloud Computing: It is been defined as *"model for enabling ubiquitous, convenient, on-demand network access to a shared pool of configurable computing resources (e.g., networks, servers, storage, applications, and services) that can be rapidly provisioned and released with minimal management or service provider interaction."* Mell and Grance (2011), where the traditional methods of accessing the internet on hardware platforms were replaced by accessing computational resources via the internet. It has three service models such as:

- Software as a Service (SaaS)
- Platform as a Service (PaaS)
- Infrastructure as a Service (IaaS)

(b) Artificial Intelligence (AI): It is a set of algorithms, procedures, and techniques which tries to simulate the human brain intelligence by a machine. The automation provided by the IoT infrastructure highly depends on machine learning (ML) which is branch of AI. Machine learning includes techniques which learns from the data which in IoT infrastructure is the data generated by the sensors.

With these supporting technologies, IoT is still dealing with a lot of issues (Čolaković and Hadžialić 2018) such as:

- Interoperability and integration: Interoperability and integration deals with the accessibility of heterogeneous devices despite manufactured by different vendors and having varied communication protocols to communicate. A survey by McKinsey analysis shows the extent of negligence towards interoperability and stated that 40% of the profit can be obtained with the integration and interoperability (Patel et al. 2018).
- Availability and reliability: (Pokorni 2019) Reliability is a metric that defines the prospect of meeting the expectations of performance standards to grant an

accurate result. With this, it poses the biggest obstacle as because of erroneous sensing, it leads to false detection which further leads to an erroneous database and enormous delays which would make IoTdevices less reliable. Availability is another metric that deals with accessibility as well as the performance of the system. This could be a big hindrance as the demand for availability of the IoT product depends upon the application it is designed for and the types of costumers using it.

- Scalability and Data storage: It is the capability of the device to adapt to the changing environment and needs which can lead to serious issues, for example IoT devices are low energy consumption devices that have limited storage capacity and are unable to execute any anti-malware and unable to store anti-malware definition.

- Management and self-configuration: Management and self-configuration is one of the biggest issues when considering IoT devices, which are connected to, network and need to auto-configure based on the value of Received Signal Strength Indicator (RSSI).

- Network performances and QOS: The network performance and the Quality of Service are both interrelated terms. In terms of IoT framework and devices, the network plays a key role as all the nodes within an IoT framework are interconnected to each other and the quality of service refers to the capability of the system to provide priority to applications to maintain consistent performance. IoT comprises of heterogeneous devices where the QoS requirements are difficult to achieve.

- Modeling and simulation: (D'Angelo et al. 2017). This problem deals with the multiple deployed IoT devices where simulator scalability issues widely affect the resultant outputs. Many hybrid simulation and modeling techniques are proposed but still, the issues of interoperability among the simulators and the designing of the inter-model interaction still exist.

- Environmental issues: This is one of the important issues in the field of IoT environment where the environmental conditions affect the sensing capability of sensor nodes, as they are sensitive to the physical changing conditions.

- Power and energy consumption: When considering the IoT devices, which are wirelessly deployed in any scenario, depending upon the application where they are used. The battery backup support is the biggest limitation.The amount of sensing leads to more consumption of energy and so it becomes a concern to be considered.

- Security and privacy: Security and privacy is the major and prominent issues of the current scenario where the entire world is heading towards the adoption of smart devices and their integration with other devices. This is leading the major issue of security and privacy of the IoT devices. Taking this aspect as the major concern, this book chapter gives insight into the major concern and measures taken for the threats that occur.

10.3   IoT Architecture and Its Security Issues

IoT framework mainly works according to its four-layered architecture as illustrated in Fig. 10.2 below, which includes the following layers:

(a) Sensing layer
(b) Network layer
(c) Service layer
(d) Application layer

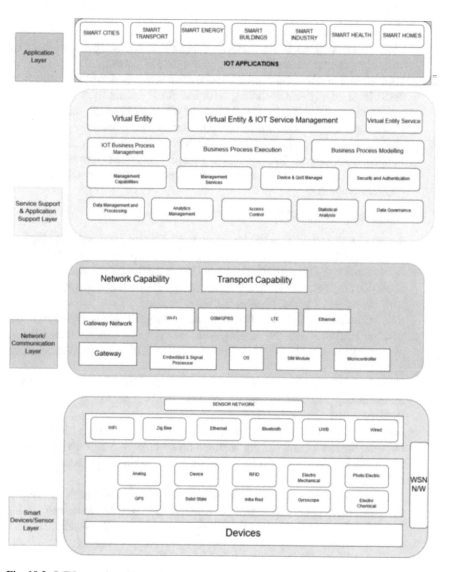

Fig. 10.2   IoT Layered Architecture

(a) **Sensing Layer**: It is the lowest layer in the IoT architecture which connects the physical world to the digital environment. It retrieves the analog data, converts it into the digital data and passes it on to the main microcontroller. It includes all the wired and wireless connections within a PAN or a LAN. The various protocols for connectivity it support are ZigBee, Bluetooth, 6Low PAN, and RFID. As IoT devices are connected as a single device within a network to the main control center without any prior supervision in specific areas, with these protocols there are various threats associated with it. The various threats are:

- Attacks on hardware: (Anderson and Kuhn 1996, 1997; Kömmerling and Kuhn 1999)

  – *Physical force*: This is the adverse category of attack specially done by humans to tamper the nodes present within the IoT network.
  – *Invasive and noninvasive attack*: Noninvasive attack is done by mishandling the voltage supply and meddling with the clock signals. Low voltages and sudden high voltage can lead to faulty operations within a microcontroller or a processor. An invasive attack is done after breaking the chip packing and altering the basic properties of the nodes by initiating a probe attack.
  – *Semi-invasive attack*: This is another category of attack, which is initiated with tools such as UV rays, X-rays, and electromagnetic radiations (Skorobogatov 2001).

- Attacks related to RFID

  – *Reverse engineering*: This can be taken up by hackers to retrieve the chip and study its working and gain access in case of RFID tags (Smiley 2016).
  – *Power Scrutiny*: This sort of attack study the power intake stages in RFID tags to decode the passcode that means to investigate the correct or wrong password with the stages of power (Smiley 2016).
  – *RFID Cloning and Spoofing*: In this attack the original data is copied from the existing RFID tag and the cloning attack is applied by using cloned tag in place of the original one (Deogirikar and Vidhate 2017; Li et al. 2012; Andrea et al. 2015).
  – *Denial of Services*: This kind of attack is initiated by using jammers, noise intervention, obstructing radio signals, and deactivating RFID tags (Smiley 2016).
  – *Eavesdropping and replay attack*: In this type of attack the information between the reader and the tag is stolen on specific protocols and the replay attack works in support of it by recording and replaying it again for unethical access (Smiley 2016).
  – *Sniffing and middleman attack*: This attack also tries to demolish the system by eavesdropping the connection information between the reader and the tag by misleading and disguise as the original component of the system.

- *RFID Skimming*: This is a threat where the attacker can skillfully steal the chip information by designing the customized device, keep it hidden below the original reader, and retrieve the confidential data.

- Attacks related to Bluetooth (UNSEF) Workshop (14 May 2015)

- *Blue Jacking Bluetooth attack*: In this the attacker can send malicious messages to a Bluetooth device, harming the device as well as the entire network associated with it.
- *Blue Snarfing Bluetooth attack*: In this attack, the attacker uses tools like hcitool, obexftp to harness information from a mobile device.
- *Blue Bugging Bluetooth attack*: In this attack the attacker deceits and takes the ownership of the mobile device and can eavesdrop the messages sent over the network.
- *Mystification*: This attack is attempted in a mysterious way by hiding the detection, in other words, it is like an ambush.
- *Surveillance*: This kind of attack has threats such as redfang, blueprinting, war nibbling, bluefish, and bluescanner in which the device is supervised to retrieve the confidential information.
- *Malware*: It is an attack where the data is stolen by introducing harmful software into a device. Attacks like CommWarrior, Skuller, Bluebag, and Carbie are popular malware to distort the Bluetooth devices.
- *PIN deducing attack*: This type of attack is targeted when the connection is to be established within two devices; the hacker uses frequency sniffer tool to device address and EDIV or RAND of the device and then brute force algorithm is applied to find all the combination of pin so as to recover the correct PIN.
- *MAC Spoofing attack*: This attack begins with the formation of a piconet within the Bluetooth topology as link keys are generated. The attacker can mimic as a user as mislead to the network operation.

(b) **Network Layer**: This layer in IoT framework allows the devices to get connected to the network and maintains interaction on the basis of the principles of data connectivity and communication. LTE, Wi-Fi, LTE-A, and Ethernet are the basic protocols that work in this later to establish connection and perform. With these connection services, the smart devices or nodes are connected to each other within a room as well as the outside world. They can pass on the messages as per the application but as it has ease of operating, it brings many vulnerabilities as well.

- *Evil Twin Hot spot*: Under this attack fake Wi-Fi spots are created by attackers to steal the information (Guest Writer 2019).
- *Packet Sniffing*: The attacker under this attack may intercept, divert, or supervise the network traffic of data packets passed over the cloud (Guest Writer 2019).

- *Malware and ransomware attack*: The IoT devices could be compromised by the injection of malware specially when connected to the public Wi-Fi hotspot (Guest Writer 2019).
- *Denial of Service*: This usually occurs because most of the devices use similar unlicensed frequency which leads to unintentional DoS (Phifer 2010).
- *War driving*: This is a kind of attack that is actually driving around and knowing what sort of wireless network exists in a specific area and what sort of encryption is been used so as to crack the passwords and publish it online on some android application or website (Joe Gray, Sword, and Shield Enterprise Security 2019).

(c) **Service Layer**: This layer in the IoT architecture is the one that performs analytics on the data collected from the nodes. The interrupted data received on to the cloud is analyzed by the enabling technologies such as artificial intelligence, machine learning, and big data. The cloud is a platform that is cost-efficient, provides flexibility, agility, responsive to fetch the results and intimate the actuator to work upon the assigned task in an IoT-enabled application. But as it is efficient to use, it is still defenseless at some issues such as:

- *Data breaching*: It is an actually unplanned proclamation of sensitive data to an unauthorized organization by unethical means caused mainly due to human error or sometimes it could be a malicious insider as well.
- *Ensure Application Program Interface*: Flaws in the APIs lead to failure of the entire system. This is because of the system or application vulnerability which can be compromised and lead to hacking of the entire network.
- *Account hijacking*: In this attack, the credential is stolen, and the cloud account is compromised to perform any malicious attack (Lord 2018).
- *Advanced persistent threat* is a kind of attack where a group attacks a network and achieves unauthorized access and remains unnoticed for a longer duration of time (Wikipedia 2019).
- *Lack of interoperability*: Cloud delivers numerous services to the users with which it can accomplish various tasks with the help of computational means which actually detach the operator with their workloads which actually leads to security gaps which in turns is the major indication of data loss (Ko and Choo 2015).
- *Abuse and nefarious use of cloud services*: To use a cloud service registration is a mandatory step, anyone possessing a credit card can access the registration and use the services of cloud but due to vulnerable registration systems malicious attackers can take over the confidential details and harm the user (Essays UK 2013).
- *Denial of service*: Attackers sometimes target a specific server and flood the data packets using botnets which actually interrupts the ongoing or initiating task.
- *Shared technology issues*: Cloud services are based on various VM and it has the feature of holding many users at a single time which makes it vulnerable sometimes because this permits the attacker to access various other VMs in

the same setup With this feature of the cloud service the user use the resourses carelessly and end up in a trap of the hacker (Baudoin et al. 2017).

(d) **Application Layer**: (Millien and George 2016) This is the last layer in the IoT architecture with the prominent interface for the user to interact with the entire IoT application in an abstract form. Depending upon the domain of the problem statement applications are designed for smart and ease of use but due to system faults, applications become vulnerable to be compromised. The various attacks used to exploit these shortcomings are:

- *Cross-Site Scripting*: It is actually an ambush containing malicious script within the page to be run on the client-side, which would manipulate the user in the desired way of the attacker.
- *SQL Injection*: It is an attack that is launched to outbreak the database of the user, in other words, it is unauthorized access to the database.
- *Insecure Cryptographic Storage:* This attack is attempted because of improper storage of data without securing it using encryption.

10.4   Solutions and Recommendations

As the IOT framework is applied in numerous domains to achieve smart, efficient and automatic solutions, its architecture needs to be strong. The prevailing threats need to be mitigated in each layer whether it is concerned with the hardware or the software. Each layer needs to use a specific defensive mechanism in order to achieve a secure network. The basic solutions for each attack are as follows:

Solutions to mitigate attack on RFID (The Government of the Hong Kong Special Administrative Region 2008)

- Password protection of the tag memory: This would help to protect or prevent the data of the tag to be unethically accessed by the attacker.
- Physical Locking of the tag memory: This measure includes the locking of the chip before releasing it into the market. The biggest limitation that comes up here are chips would only have read mode, they are not enabled to be editted.
- Reader's Protection: To protect the tag information the author encrypts it with the private key and the reader then gives its information as the public key to decrypt and use the tag.
- Read detector: If anyone tries to modify or kill the tag, a message could be passed over reticent frequency describing the unethical attack on the tag.
- Faraday Cage: Tags are secured in a metal covering, which is known as a faraday cage, which prevents the frequency and protects the tagged objects to be detected.
- Jammers: These are the obstruction, which jam the neighboring RFID reader to read the tags.

Solutions to mitigate attack on Bluetooth (Minar and Tarique 2012):

- Pin selected for the connection orientation in a Bluetooth needs to be long comprising of capital letters, small letters, special characters, and numerals.
- Bluetooth configuration should be configured as undiscoverable so as to avert the visibility to other Bluetooth devices.
- The transmission of the data through the Bluetooth connection is vulnerable to eavesdropping and to protect it, encrypt the connection to make it secure.
- To establish a secured link-connection security mode 3 needs to be provided and to have a secure link-level connection authentication and encryption with mode 2 and 4 should be implemented.

Solutions to alleviate attack on network layer of IoT

(a) Resolutions to prevent evil twin hot spot: To maintain prevention, both the perspectives and activities of the customers and owners providing the services of Wi-Fi (Hunt 2018) need to introspect and perform the following:

- Being a customer, following points need to be kept in consideration while connecting to any free Wi-Fi zone.

  – Confirm the official Wi-Fi name before connecting to the available hot spots. To test twin hot spot, deliberately try to use the wrong key, if some keys govern the hotspot.
  – If the wrong key is accepted by the system, then it is definitely a twin hotspot case.
  – Always disable the automatic connection options for the saved hotspots, as they are the vulnerable spots for the twin hot spot.
  – The devices should be disconnected after a certain period and reinitiate the connection manually to an authentic hot spot.

- Being a proprietor or service provider performs the following activities to protect the network from malicious events of people.

  – Proclaim the name of the network evidently despite creating a Wi-Fi and distributing the key to the customers.
  – Try to investigate the network with any network-connecting gadget and alerting the consumers about caricaturing of the official network.
  – If someone has enduringly positioned an evil twin on your network, then hire a professional to track the position of the malicious access point. Applications such as EvilAP Defender should be installed which notify the existing evil twin.

(b) Resolution to mitigate the sniffing attack (Passi 2018): The various steps are:

- Connection to an authentic network: To prevent sniffing attacks avoid connecting to the public networks as the attacker could impersonate the authentic network in a honey trap of free service.
- Encryption: A strong encryption mechanism should be implemented to the data packets leaving the system and moves to the server or the cloud and back from the smart gadgets to the cloud and the actuator to prevent the data breaching.

- Network Monitoring: Network supervision is required to detect any intrusion that would capture the moving data traffic.

(c) Solutions to mitigate the attacks of malware and ransomware attack: Ransomware is one of the malware attack that is very dangerous for IoT devices as they are vulnerable with the situations where a hacker may take access of devices connected to the main power supply, which could lead to serious issues if ransom not paid back. To avoid such situations, following steps need to be taken into consideration (Koopman 2017):

- Passwords: The manufacturer should set that; the default passwords of the remotely placed devices should alter every time before deploying them into such an environment.
- Customer access to alter password: If the device is deployed in remote, the device should switch off its main function and enable the customer to change the password.
- Updates: The devices regularly are updated, and the devices should automatically install them.
- Reset function: The reset function of the remotely placed devices should not reset by anyone except the manufacturer.

(d) Solution to prevent the devices from DoS attacks: Keeping IoT in consideration is one of the most impactful attack. IoT comprises heterogeneous devices where DDoS focuses on targeting one node within a network with multiple connections. To prevent the network from a DDoS, following measures needs to be taken (Rubens 2018):

- More bandwidth: This involves strengthening the network by improving the resistance in terms of increasing the bandwidth of the network. In order to launch a DDoS attack, it is hard to buy more bandwidth.
- Akamai DDoS mitigation: It is content delivery network-based, consumer's requisite based. It alleviates DNS-based DDoS attacks such as DNS amplification, as well as protecting DNS services. It monitors tools, data traffic strategies, and web application firewalls (WAF) rules.
- Structuring redundant infrastructure: To protect the DDoS attack distribute the data traffic between the various data centers, separated geographically or topographically. With this distribution, it is hard for the attacker to damage the servers.
- Verisign DDoS Protection Services: It monitors and observes any DDoS attack and notify the consumer and even proposes a strategy to fight against it. It has an API known as Open Hybrid, which empowers the security system of the organization to send the threat information and the immediate solution for the same.
- Deployment of DDoS resistant hardware and software: To protect the network as well the entire security system of any organization specialized configured network nodes must be installed and deployed which may block certain ICMP data packets or DNS requests which are from the outer

network. Software protection is also provided by protecting servers using network firewalls and web application firewalls which may protect the network from SYN flood attack by observing the incomplete connection within the network and discarding the one when it reaches a threshold value.

- Radware DDoS Protection: It is an integrated web application as well as a network security solution.
- Cloud flare DDoS Protection: It deals with providing solutions to cloud-based attacks for layers 7, 3, and 4 because in DDoS attack every network node gets involved in the attack.
- Arbor Networks APS: It is made use of multi-layer and hybrid fortifications to prevent the system from all sorts of DDoS attacks. In terms of physical protection, it is provided by Arbor's APS that reports about TCP state-exhaustion attacks in the physical layer.
- Nexus guard: It provides a way out for all types of DDoS attacks and cyber threats. This encompasses protection against level 3 to level 7 attacks, including DDoS attacks, brute force, connection flood, ping of death, Smurf, SSL flood, zero-day attacks, and more.
- DoS arrest DDoS Protection: The DoS arrest attack mainly targets mobile TCP ports such as 80 & 443, HTTP websites. The DoS arrest and DDoS protection is an integrated cloud-based security that uses big data analytics engine, which provides prevention against DDoS attack, monitors website, and provides a web application firewall.

(e) Solution to prevent IoT devices from wardriving: Wardriving is a particular kind of piggybacking attack which can be mitigated by following steps such as:

- Mandatory changing of Default passwords: The access points have default passwords set by the manufacturer, so it is mandatory for the user to change the password and continue to update it periodically.
- Restriction on the accessibility: Accessibility is one of the issues, which needs to consider on top priority by restricting the users based on the MAC address by filtering it. Creating a special guest account to grant access to any guest on a discrete wireless channel with a separate password aids in maintaining the privacy of the primary passwords.
- Encryption of data: Data flowing to and fro in an encrypted form protects it from hampering or breaching protocols, for example WEP, WAP, WAP2, and WAP 3 can be used to encrypt the data.
- Maintenance of SSID protection: SSID stands for Service Set Identifier, which uniquely identifies a particular Wi-Fi network from other networks. To prevent the network, do not publish SSID. Change the default SSID to a distinctive one in order to protect it from the hacker.

Solution to avert Service layer from malicious attacks:

- Assurance of operational authority for risk and defiance processes: Every organization defines its own regulations, privacy and security policies based on the analysis of the IT assets a company possesses. A good operational authority through its policies defines the roles and responsibilities of the people working in the firm as well as the customers connected to them and specifies transparency in terms of business and legal issues. Services provided by the cloud are IaaS, PaaS, or SaaS, which has a substantial influence in distributing the accountabilities among the consumers and the service providers to achieve safety and mitigate the allied menaces.

- Auditing of the processes: The auditing is a very important process performed by the IT companies, which confirms their acquiescence with the policies and terms of government, which should be visible to the consumers. Here there are crucial areas of security which is considerate to both the customers as well as auditors such as:
- Guaranteeing seclusion of the client's data and application in a shared environment. Protecting the consumers' possessions from malicious attacks, accidents, or intended access.
- For analysis of how any suspicious activity took place, there is a need for widespread vision of security controls which comprises both data flow and the consumer's application from service provider to the customer.
- Managing roles, responsibilities, and identities: The service providers should make sure they abide by the rules and regulations of the policies as given by the government, provide and manage distinctive identities to their clients and services which is known as identity and access management (IdAM). Identity management provides access privilege to all clients, administrators, developers, and testers by the concept of key management such as cryptographic keys, passwords, and Multi-Factor Authentication (MFA).
- Protection of data: Data is the most crucial asset that has to be protected from any attack. Data is in two forms such as data, which is stored and another form is the data that is in transition from one place to another. Cloud being distributive in nature, securing the data from breaching or hampering is the most challenging issue. For such an issue, numerous controls are adopted such as:
  - Creating a data asset catalog.
  - Organizing all forms of data.
  - Applying all privacy concerns such as passwords.
  - Applying the CIA principles onto data.
  - Permitting data monitoring.
  - Security logging and checking.
  - Evaluate the security necessities for cloud applications: To evaluate this focus needs to be on the coding which comprises of following steps which need attention:
  - Applying appropriate endpoint protection and API security.
  - Input validation.
  - Output encoding.
  - Session supervision.
  - Password Management.
  - Safety of profound data in storage and in transition.

- Error management and logging.
- Security of log data.
- Assortment and efficient usage of APIs and network amenities.

Solutions to prevent application layer from cyber attacks

1. Steps to prevent Cross-Site Scripting (Ports Wigger: Web Security 2019): To prevent cross-site scripting, following steps needs to be performed:

   (a) Clean the input data on advent: When the user filter receives input check, the input received as per the expectation to achieve a valid input.
   (b) Encoding of Output data: To prevent the data from being understood, the output data is required to be encoded.
   (c) Apt usage of response headers. To avoid XSS in HTTP responses that do not have HTML or JavaScript, Content-Type and X-Content-Type-Options headers can be used to confirm that the browser can interpret the response according to the developer.
   (d) Content Security Policy. Content Security Policy (CSP) is used to decrease the harshness of XSS susceptibilities that arise.

2. *SQL Injection*: To prevent the system from SQL injection, following recommendation is necessary to consider:

   (a) Validation: User input data may have some bugs, to remove and validate it via function such as MySQL's mysql_real_escape_string() and filter the content.
   (b) Avoid the usage of dynamic SQL: Never create queries with the help of user input data instead use constructs defined such as parameterized queries or stored procedures but do not rely completely on these constructs as they are susceptible to SQL injection attack.
   (c) Regular Updating and patching: Continuous monitoring of the vulnerabilities in applications and databases that malicious people can develop using SQL injection is required. The applications should be regularly updated on a periodic basis.
   (d) Firewall: A web application firewall should be there within the system to filter out malevolent data. It is beneficial in a way that provides protection to the data even when a patch is not available. A free open source is available for web browsers such as Apache, Microsoft IIS, and nginx which provides security to malicious SQL injection.
   (e) Usage of suitable privileges: Providing limited privileges to access any account can always help protect the system from the hacker to perform any spiteful activity and never connect any user to the database using an account with admin privilege, for example considering a web login page should have connection with account which has limited scope to credential table.
   (f) Periodic monitoring of SQL queries: Monitoring of the SQL commands is necessary for database connected application.

10.5 Real Attack on IOT Devices: A Case Study

1. *Attacks using DoS* (GlobalSign Blog 2016): In *2013*, it was observed that over 39 attacks were attempted above 100 gbps (gigabits per second), which has been increasing ever since. In the month of march "the spamhaus ddos attack", 120 gbps of traffic got hit the network, another in august 2013, another DoS attack was noticeable when partially the "chinese internet system broke down" and the government was not able to defend the attack.

   In the year 2015, the world-renowned "The New York Magazine" got knockout by a DDoS attack the moment it publishes the interview session of 35 women accusing Bill Cosby for sexual assault. Another incident within the same year was when a UK-based car phone Warehouse got under attack where the hacker stole consumer's personal data using a DDoS attack. By the end of the year 2015, Microsoft's Xbox Live and PlayStation network was on complete shutdown for a week due to a DDoS attack where attackers attempted to show the weakness of the Microsoft services.

   In the year *2016*, it was one of the largest attacks of DDoS where the consumers of HSBC bank lost their authenticity to access their own accounts for online transaction just before the last date of the payment of the tax in the UK.

   In the year *2018*, GitHub was under the DDoS attack which broke all the above records of 1.3 terabits per second where the technique known as mem caching was used as an attack that is a database caching system used to accelerate websites and networks. The attackers managed to spoof GitHub's IP address and then tremendously intensified the levels of traffic (Strawbridge 2019).

   According to digital trends report the DDoS attack has grown by 132% compared to the year 2014 and the rate of empolying IOT devices as well, which can lead to further attacks in near future.

2. *Attacks using botnet on smart home appliances*: Proofpoint, a California-based research group, observed that cyber crooks would target the smart devices and domestic routers and transmute them into "Thingsbot" which would perform the attack. The attack they observed was from December 23, 2013, to January 6, 2014, where a campaign for the global attack was performed by compromising a smart fridge which included 750,000 malicious emails from 100,000 devices three times a day.

3. *Attack using middleman concept*: Charlie Miller and Chris Valasek the cybersecurity expert demonstrated, how they broke into the security system of a car named Jeep Cherokee. They were able to exploit the vulnerable element named Uconnect, which is internet connect computer feature in almost all the cars of Fiat-Chrysler cars, SUVs, and trucks. Uconnect offers a Wi-Fi hotspot facility, manages and controls the navigation and entertainment of the vehicle. Both the specialist made Uconnect as an entry point in the head unit of the vehicle and instead put their own code and rewrote the chip. This changed the firmware and the access to CAN bus was achieved which ultimately gave the entire control in the hands of the hacker.

10.6   Conclusion

With the upcoming changing trends of the market, numerous new sort of smart devices will be introduced , which could be more vulnerable. As per the growing demands, the technologies should blend together to have innovative solutions to maintain the security system. The enabling technologies such as machine learning, artificial intelligence, big data, and many more combinations can be built up together to build up security system such as IDS. IDS stands for Intrusion Detection System, which could be an automatic system detecting any sort of intrusion and provide security at all the layers and make this planet a better and smart place to live.

References

Published Proceedings

I.F. Akyildiz, W. Su, Y. Sankarasubramaniam, E. Cayirci, Wireless sensor networks: A survey. Comput. Netw. **38**(4), 393–422 (2002)

R. Anderson, M. Kuhn, Tamper resistance—a cautionary note, in *Proceedings of the Second Usenix Workshop on Electronic Commerce*, vol. 2, (1996), pp. 1–11

R. Anderson, M. Kuhn, Low cost attacks on tamper resistant devices, in *International Workshop on Security Protocols*, (Springer, Berlin, Heidelberg, 1997), pp. 125–136

I. Andrea, C. Chrysostomou, G. Hadjichristofi, Internet of things: Security vulnerabilities and challenges, in *2015 IEEE Symposium on Computers and Communication (ISCC)*, (IEEE, New York, 2015), pp. 180–187

C. Baudoin, E. Cohen, C. Dotson, J. Gershater, D. Harris, S. Iyer, *Security for Cloud Computing Ten Steps to Ensure Success Version 3* (Cloud Standards Customers Council, Needham, MA, 2017)

A. Čolaković, M. Hadžialić, Internet of things (IoT): A review of enabling technologies, challenges, and open research issues. Comput. Netw. **144**, 17–39 (2018)

G. D'Angelo, S. Ferretti, V. Ghini, Modeling the internet of things: A simulation perspective, in *2017 International Conference on High Performance Computing & Simulation (HPCS)*, (IEEE, New York, 2017), pp. 18–27

J. Deogirikar, A. Vidhate, Security attacks in IoT: A survey, in *2017 International Conference on I-SMAC (IoT in Social, Mobile, Analytics and Cloud) (I-SMAC)*, (IEEE, New York, 2017), pp. 32–37

Essays, UK: Abuse and Nefarious use of Cloud Computing Information Technology Essay (2013), https://www.uniassignment.com/essay-samples/information-technology/abuse-and-nefarious-use-of-cloud-computing-information-technology-essay.php?vref=1

R. Ko, R. Choo, *The Cloud Security Ecosystem: Technical, Legal, Business and Management Issues* (Syngress, Boston, 2015)

O. Kömmerling, M.G. Kuhn, Design principles for tamper-resistant smartcard processors. Smartcard **99**, 9–20 (1999)

M. Koopman, Preventing Ransomware on the Internet of Things (2017)

E.A. Kosmatos, N.D. Tselikas, A.C. Boucouvalas, Integrating RFIDs and smart objects into a Unified Internet of things architecture. Adv. Internet Things **1**(01), 5 (2011)

H. Li, Y. Chen, Z. He, The survey of RFID attacks and defenses, in *2012 8th International Conference on Wireless Communications, Networking and Mobile Computing*, (2012), pp. 1–4

P. Mell, T. Grance, *The NIST Definition of Cloud Computing. Special Publication 800-145* (US Department of Commerce, Gaithersburg, MD, 2011)

N.B.N.I. Minar, M. Tarique, Bluetooth security threats and solutions: A survey. Int. J. Distrib. Parallel Syst. **3**(1), 127 (2012)

G. Nunberg, The Advent of the Internet (2012)

D.N. Patel, D. Yadav, B. Morkos, Analysis of consumer response and pricing of smart and connected products, in *ASME 2018 International Design Engineering Technical Conferences and Computers and Information in Engineering Conference*, (American Society of Mechanical Engineers Digital Collection, 2018)

S.J. Pokorni, Reliability and availability of the internet of things. Vojnotehnički glasnik **67**(3), 588–600 (2019)

UNSEF: "A Brief History of the IoT" United Nations Social Enterprise Facility (UNSEF) Workshop on Internet of Things Development for the Promotion of Information Economy Boracay, Philippines 14 May 2015 (2015)

Web References

GlobalSign Blog: Closed for Business—the impact of denial of service attacks in the IoT (2016), https://www.globalsign.com/en/blog/denial-of-service-in-the-iot/

Guest Writer: Securineg an IoT Solution? Start with the Network Layer (2019), https://www.webtitan.com

G. Hunt, How to detect and escape Evil Twin Wi-Fi Access Points (2018), https://www.titanhq.com/blog/how-to-detect-and-escape-evil-twin-wi-fi-access-points/

Joe Gray, Sword & Shield Enterprise Security: Wireless network and Wi-Fi security issues to look out for in 2019 (2019), https://www.alienvault.com/blogs/security-essentials/security-issues-of-wifi-how-it-works

N. Lord, What is Cloud Account Hijacking? (2018), https://digitalguardian.com/blog/what-cloud-account-hijacking

R. Millien, C. George, The enabling technologies of the Internet of Things (2016), https://www.ipwatchdog.com/2016/11/28/enabling-technologies-internet-things/id=75039/

H. Passi, What is a Sniffing attack and How can you defend it? (2018), https://www.greycampus.com/information-security/what-is-a-sniffing-attack-and-how-can-you-defend-it

L. Phifer, Top ten Wi-Fi security threats (2010), https://www.esecurityplanet.com/views/article.php/3869221/Top-Ten-WiFi-Security-Threats.htm

Ports Wigger: Web Security: Cross-site scripting (2019), https://portswigger.net/web-security/cross-site-scripting

P. Rubens, How to prevent DDoS attacks: 6 tips to keep your website safe (2018), https://www.esecurityplanet.com/network-security/how-to-prevent-ddos-attacks.html

S.P. Skorobogatov, Semi-invasive attacks (2001), https://www.cl.cam.ac.uk/~sps32/semi-inv_def.html

S. Smiley, 7 types of security attacks on RFID systems (2016), https://blog.atlasrfidstore.com/7-types-security-attacks-rfid-systems

G. Strawbridge, 10 biggest DDoS attacks and how your organisation can learn from them (2019), https://www.metacompliance.com/blog/10-biggest-ddos-attacks-and-how-your-organisation-can-learn-from-them/

The Government of the Hong Kong Special Administrative Region: RFID security (2008), https://www.infosec.gov.hk/english/technical/files/rfid.pdf

Wikipedia: Advanced persistent threat (2019), https://en.wikipedia.org/wiki/Advanced_persistent_threat

Chapter 11
Authentication Attacks

Ankit Vishnoi

11.1 Introduction

Authentication is procedure of approving the user's identity. Clients are distinguished utilizing diverse authentication systems. In a security framework, the authentication procedure checks the data given by the client with the database. If the data matches with the database data, the client is conceded access to the security framework. A wide range of authentication components permits client to gain admittance to the framework, anyway they all work in an unexpected way. Any user is allowed to log into protected sites through a set of access control mechanisms. Each access control mechanism has four procedures which are identification, authentication, authorization, and accountability.

The identification is the point at which the client enters the ID and ID is checked with the security framework. Some security frameworks create random IDs to secure against the cyber attackers. There are three authentication forms. Authorization is checking and coordinating the authenticated element of data with access level. The authorization procedure is handled in three different ways—authorization is performed for authenticated client, authorization is performed for members of a particular group, authorization is performed over the various frameworks, and accountability is a procedure keeping framework logs. Framework logs monitor all fruitful and ineffective logins.

A. Vishnoi (✉)
School of Computer Science, University of Petroleum and Energy Studies, Dehradun, India

© The Author(s), under exclusive license to Springer Nature Switzerland AG 2021   217
A. Bhardwaj, V. Sapra (eds.), *Security Incidents & Response Against Cyber Attacks*,
EAI/Springer Innovations in Communication and Computing,
https://doi.org/10.1007/978-3-030-69174-5_11

11.2   Authentication Process

The process of proving an assertion like the identity of a user towards a computer system is called authentication. The mechanism of authentication mainly comprises of association of an incoming request with a set of identifying credentials in the database of the system. The user is first required to submit his/her credentials to the system database by creating a user ID and a key with which the user can be authenticated and authorized into the system later. It is important to note that authentication is a process of verifying identity of the user whereas authorization dictates granting access of specific resources of the system/website to the authenticated user (Kraus et al. 2016). The authentication process is the foremost step that takes place in any protected system before any other function can begin initiation.

The process of authentication has three tasks:

- Direct connection establishment between user and the system and website server and its management.
- Verification of the user identity.
- Providing approval (or denial) to user authentication so that system can move to authorizing the user.

Authentication of a user is crucial and needs to be a very secure procedure because, if otherwise, it could lead to data breaches. Restricting an unauthorized user from entering the system and gaining access to sensitive information and providing an authorized user only the resources and information he is liable to access are two utmost important reasons for building a highly secure authentication system.

The mechanism of authentication can be described in five simple phases:

- User input of his/her credentials in the form of username and key in the login form.
- The information provided by the user is sent to the authentication server.
- The user credentials are then compared to those in the database of authorized user's information.
- If a match is found, the user is authenticated and granted access to his/her account.
- If a match is not found, the user is prompted to re-enter the required details and try again.

The term *"key"* repeatedly being mentioned in the description of *authentication* refers to nothing but an *authentication factor,* a piece of information provided by the user that has to be restricted to the knowledge of the user and of the server to avoid data breaches or cyber-crimes.

There are mainly five types of authentication factors:

- **Something you know (Knowledge-based factors)**: A piece of information which is stored in user's memory and can be retrieved when needed is classified as a knowledge-based factor. It can be anything, a password, a PIN (personal identification number), a security question's answer, etc. In comparison with other factors, this one is considered to be on the weaker side in terms of security due to the risk of it being guessed or shared to someone capable of misusing it. Knowledge-based factors can be used to perform two kinds of authentication:

- Static knowledge-based authentication (SKBA): This is normally used as a backup authentication way for the users by banks, e-mail providers, or financial services companies in case the user forgets his/her password. It has proven to be the easiest way to prove the identity of the user and grant access of the account without hustle usually with the help of secret security questions.
- Dynamic knowledge-based authentication (DKBA): DKBA provides higher level of security by requiring the users to answer randomly generated questions instead of preset security questions based on the information from user's personal data file in the database, credit reports, or the user's marketing data.

- **Something you have (Possession/Token-based factors):** Information that can be physically carried by the user and be used to authenticate his/her identity is called as a possession/token factor. One-time passwords (OTP), ID cards, physical tokens, smart cards, and key fobs are all examples of possession factors. OTPs are usually 6–8 digit long that expire either after the set time limit by the OTP generation server or after the first use. Smart cards are used in high security offices like federal offices to grant employees access to facilities and systems. Possession factors are often combined with knowledge-based factors to form a two-factor or dual factor authentication system.
- **Something you are (Biometric-based factors):** Any physical part of the human body that can act as information which can be offered for verification comes under biometric-based factors. Some of the biometric authentication process performed using these factors include fingerprint scanning, thumbprint scanning, facial recognition, retina scanning, voice verification, etc. Although the biometric factors provide highest level of security among all factors as they are extremely difficult to forge but they have one fundamental flaw: digital signatures are used for the process of recognition, which, like any other information, can be hacked if kept unprotected and fed to the website/system.
- **Somewhere you are (Location-based factors):** This factor is not as commonly used as the ones mentioned earlier. Location detection technologies like the IP (internet protocol) address and MAC (media access control) address are used to validate the authenticity of the user or to notify the user about an unusual access attempt. Services like ATMs, e-mail logins use geolocation security checks to track access attempts of users. Attempt to log in into the account from a new IP address notifies the customer about the activity and often confirms the user's identity. Debit/credit cards being used from two different geolocations can lock down account till user authentication completion to avoid thefts. MAC addresses are highly useful when organizations restrict entry into their servers to some specific MAC addresses.
- **Something you do (Action/behavioral biometrics-based factors):** This factor is based on the question "*what makes a person unique?*" Yes, the biometrics but

as mentioned earlier the digital signatures can be stolen. It is the person's behavioral characteristics. No two people can have the same behavioral characteristics. What is the first thing that the user accesses on the phone in the morning, the song that is played by the user every single day, people contacted by the user frequently, the most visited website, etc. These are things, which can never be duplicated by someone else. The machine being used by the user (phone, laptop, tablet, etc.) can track these activities in real time and analyze them to verify your identity on an ongoing basis. This technology works on the principles of machine learning and is the most powerful factor of authentication available.

11.3   Existing Authentication Methods

11.3.1   Static Authentication by a Password

This type of authentication is widely known for its simple implementation. The user is required to recall a piece of information submitted to the server database as the key or password to a specific user ID. A password can be any of the following things:

- A character password
- A numeric password/ passcode
- A passphrase (is longer than a password and maximizes on key space)

A solid strong password is considered the one, which is a combination of one or more upper case letters, lower case letters, special characters, numeric characters and has minimum length of 12 characters. If static authentication is not used carefully, then it can be considered as a low security solution and has many drawbacks which include eavesdropping of the password, easy physical attacks like password being recorded through a camera when it is being typed, vulnerability of static passwords to guesswork or brute force attacks is very high, risk of a replay attack which is actually a sort of deception/impersonation that involves using information from single previous execution of a protocol on a verifier (which maybe same or different from the last time).

Among other issues, these requirements were addressed with productivity in mind:

- Web-based application for registration, approval, and establishment of account.
- Web-based administration of accounts by delegation, based on distributive rights and obligation.
- Self-service treatment of editable personal data and changing of password.
- Disallow the use of special local alphabetic characters in passwords, such as Å, å, Ø, æ, Ø, and ø, which will eventually cause authentication issues due to different types of keyboards.

Implementation of security policies is the major undertaking, with a few relevant implementations being:

- One person's stance on one account. Just two people or more that share an account. One individual possesses just one user credential.
- Automatic user identifiers would ensure consistency and security against trivial mapping of user name to user id, e.g., a "pew028" identifier will be assigned to "PeterWhite" instead of "peter."
- Use strict authentication rules, e.g., more than 10 characters long, combine upper case, lower case, digits, and non-alpha numeric for password.
- Enforce outdated login and recycle vs. policies with password. Implement automatic account expiry and deletion based on recorded policies, which are publicly accessible.

Drawbacks of Username/Password:

- Passwords are difficult to search but simple to recollect.
- Passwords are handily taken whenever recorded.
- Clients can share passwords.
- Passwords can be overlooked if hard to remember.

11.3.2 One-Time Password (OTP)

OTPs serve the purpose of annihilating password replay-ability by generation of new password every time. OTP systems are usually combined with static authentication to form a two-factor authentication system. They act as an extra layer of security to all authenticated applications as even if the static password is compromised somehow, breaking of OTP will be a bigger task for any cybercriminal to unethically gain access. OTPs do not facilitate any changes in the internal system or database and only impact the monitors (Huang et al. (2013), Idrus et al. (2013), Kumar and Murugan (2018), Manjunath et al. (2015), Tumin and Encheva (2012)).

Static OTPs: It is a randomly generated pin, which is valid through a single login session or a single transaction through a digital device (a computer, mobile phone, etc.). When a password is generated by the OTP system, synchronization between the OTP token and the monitor is required by the verification process. Generation of OTPs is done through algorithms based on randomness or pseudo-randomness. The randomness of algorithms is of high priority to make it impossible for anyone to predict OTPs by observing previously generated OTPs.

OTPs can be generated by following various approaches, some of which are as follows:

1. Establishing time-synchronization between the user providing the password and the authentication server, in this process OTPs are made valid only for a limited period of time.
2. Generating a new password every time by using the previously generated password through a mathematical algorithm. Here OTPs generated form a sort of chain and should use in the predefined order.

3. Generating a new password with the help of a randomly chosen number by the authentication server, which acts as a challenge, through a mathematical algorithm as illustrated in Fig. 11.1 below.

11.3.3 Biometrics

Biometrics are quantifiable characteristics and measurable examinations of a person in perspective on their exceptional physical or behavioral components. Biometrics is these days utilized as a security component for confirmation and provision of access to programmed and instant verification based frameworks. There does not

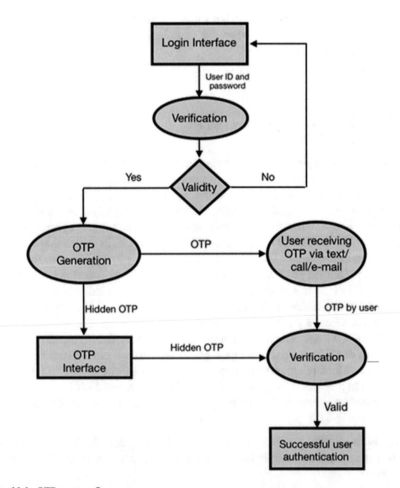

Fig. 11.1 OTP process flow

exist one "best" strategy for biometric data assembling and perusing that can ensure secure validation.

There are five kinds of biometric technologies ordinarily utilized:

1. Retina scan: The uniqueness of the pattern of the veins at the rear of the eye makes it difficult to imitate a retina. The novel and perpetual nature of the retina makes it the most exact and dependable biometric.
2. Fingerprint scan: Fingerprints of no two individuals is ever the equivalent and they stay unaltered all life aside from in instances of wounds like bruises and cuts on the fingertips of the individual. Fingerprint recognition is relatively more affordable than the retina scan.
3. Facial recognition: It is one the most versatile and working biometric identification techniques. In many cases, even the subject is unaware of being scanned or filtered. The uniqueness of each individual can be perceived in a brief instant by biometric affirmation scanners by analyzing the facial components (like width of the nose, separation between the eyebrows, arrangement of the cheekbones, and so on) in detail.
4. Voice Recognition: Voice biometrics can likewise be utilized to verify an individual's character without his/her insight. Elements like the pitch, voice level; contrasts in each individual voice are checked, experienced and approved to give the access to the user.
5. Hand/Palm Scan: This not only gives the finger tips but also the shape and size of a person's hands, the weight with which the palm was put on the scanner, the indents, and so on are additionally monitored which are utilized while comparing different palms with each other (Kraus et al. 2016).

The strong relationship present between the user and the biometric data (authenticating information) acts as the biggest advantage of biometric-based authentication methods. Biometric information along with its reference is fed to the system every time the user successfully attempts to authenticate himself. This information can be intercepted by an attacker and replayed to gain access. If the information were of dynamic nature rather than being static, it would be difficult for the attacker to intercept. Dynamic biometrics can be achieved by implementing a challenge-based password generation scheme, by establishing an OTP-based authentication system, etc. as illustrated in Fig. 11.2 below.

The process of biometric authentication is described in two steps:

1. Enrolment: This stage can be further divided into two parts. First part of the step is performed by the user, being provision of the biometric information by the user to the system. Second step involves work of the system, biometric data is captured, and its features are extracted and then stored into the user database.
2. Authentication: A comparison between the stored features with the ones already present is initiated. If a match is found, then user is authenticated and access is granted to the user.

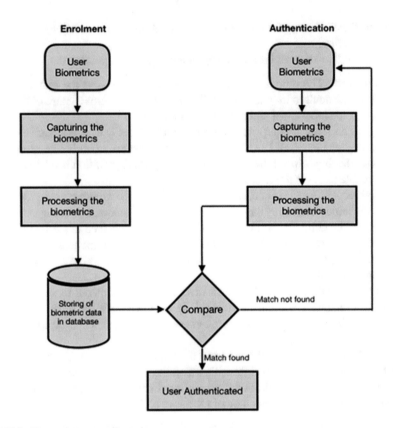

Fig. 11.2 Biometric process flow

Advantages of Biometrics:

- Biometric is one of a kind and is straightforward.
- Very hard to reproduce biometric highlight.
- Biometric attributes can't be lost.
- Biometric is utilized at significant places, for example, at air terminals, settlements reason and at detainment facilities.
- Fingerprint examine is little and modest.
- Can be utilized via telephone lines.
- Eye examine are exactness in distinguishing clients.

11.4 Execution Guidelines Solutions to Delicate Information Sensitive Information

It applies to any data that is utilized for verification, or might be utilized to the weakness of the program, gadget, or condition whenever traded off. Appropriate treatment of touchy information can expediently increase or lessen an application's capacity. Touchy information—which may incorporate passwords, keys, or other individual data—can be promptly utilized and get unmanageable all through an application's internals. There are different inside angles to that:

- Data storage and adjustment in memory
- Persistent storage
- Sending the message—IPC, RPC, organize contact

To control information items, structures, and factors managing delicate data, the thumb rule is to disconnect and confine the utilization of the information and clear it when activities are finished. Detachment is frequently progressively hard to accomplish with current memory the board; however, it could be practiced by independent procedures and strings, as is regularly found in utilization of benefit partition. Restricting the utilization of touchy data implies just working for whatever length of time that important to hold it around. The most widely recognized model is the putting away of clear-content passwords and variable certifications. This information is frequently changed before utilizing into some other structure, for example, a hash or figure content. The memory ought to be cleared (focused) and discharged once the change is finished. In those dialects, which treat strings as changeless articles, care ought to be taken rather to utilize non-permanent supports and exhibits for capacity.

References

Y. Huang, H. Zheng, H. Zhao, X. Lai, A new One-time Password Method, in *2013 International Conference on Electronic Engineering and Computer Science* (2013)

S.Z.S. Idrus, E. Cherrier, C. Rosenberger, J.-J. Schwartzmann, A review on authentication methods. Aust. J. Basic Appl. Sci. **7**(5), 95–107 (2013)

L. Kraus, J.-N. Antons, F. Kaiser, S. Moller, *User experience in authentication research: A survey* (Quality and Usability Lab, Telekom Innovation Laboratories, TU, Berlin, 2016)

K.M. Kumar, G.B. Murugan, Comparative study on One Time Password Algorithms. Int. J. Comput. Sci. Mob. Comput. **7**, 37–52 (2018)

D. Manjunath, A.S. Nagesh, M.P. Sathyajeet, N. Kumar, S. Akram, A survey on knowledge based authentication. Int. J. Emerg. Technol. Innov. Res. **2**, 1194–1201 (2015)

S. Tumin, S. Encheva, A closer look at authentication and authorization mechanisms for web-based applications, in *5th World Congress: Applied Computing Conference* (2012)

Index

Printed in the United States
by Baker & Taylor Publisher Services